Astronomers' Universe

The Astronomers' Universe series attracts scientifically curious readers with a passion for astronomy and its related fields. In this series, you will venture beyond the basics to gain a deeper understanding of the cosmos—all from the comfort of your chair.

Our books cover any and all topics related to the scientific study of the Universe and our place in it, exploring discoveries and theories in areas ranging from cosmology and astrophysics to planetary science and astrobiology.

This series bridges the gap between very basic popular science books and higher-level textbooks, providing rigorous, yet digestible forays for the intrepid lay reader. It goes beyond a beginner's level, introducing you to more complex concepts that will expand your knowledge of the cosmos. The books are written in a didactic and descriptive style, including basic mathematics where necessary.

More information about this series at http://www.springer.com/series/6960

Wolfgang Kapferer

The Mystery of Dark Matter

In Search of the Invisible

Wolfgang Kapferer
Telfs, Austria

ISSN 1614-659X ISSN 2197-6651 (electronic)
Astronomers' Universe
ISBN 978-3-662-62201-8 ISBN 978-3-662-62202-5 (eBook)
https://doi.org/10.1007/978-3-662-62202-5

The translation was done with the help of artificial intelligence (machine translation by the service DeepL.com). A subsequent human revision was done primarily in terms of content.

Responsible Editor: Lisa Edelhaeuser
This Springer imprint is published by the registered company Springer-Verlag GmbH, DE part of Springer Nature.
The registered company address is: Heidelberger Platz 3, 14197 Berlin, Germany

Foreword

Today, the concept of "dark matter" has undergone an exciting two and a half centuries of history. From initially postulated dark stars to understand the strange movements of some luminaries on the firmament, to hypothetical planets in the outer reaches of our solar system, time and again it was possible to wrest matter from its darkness and, with advancing observation techniques, to discover its true nature.

At the beginning of the twentieth century however, a large gap regarding the masses of large-scale structures in the universe derived by various methods was discovered. The larger the structures, the greater the difference between masses derived from direct observations and masses derived from the dynamics of these objects. The successful concept—dark matter—was and is an approach to resolve this contradiction. It calls for an all-encompassing, purely gravitationally interacting substance in the universe, which is otherwise—at least until now—completely invisible. This concept of hidden matter that only reveals itself to us indirectly was and is very successful. A well known example of this is the theory of the formation and development of large-scale structures in the cosmos.

To the layman however, this concept often seems erratic and arbitrary. The aim of this book is, on the one hand, to show the great historical successes of the concept of dark matter in astronomy. The most important key observations in this field of research and their interpretations will be presented. These numerous observations have led to a world view in which our universe is mainly dominated by a form of matter as yet unknown to us. On the other hand, the potential candidates for dark matter are also presented, and the reader is familiarized with the great experiments in this highly topical field of research, the search for them and the most promising results to date.

The search for the nature of dark matter resembles a quest of the grail, transported into the modern age of the natural sciences. The book cover is dedicated to this parable. It shows the "artistic" visualization of a simulated galaxy rotation curve, which resembles a chalice when it is mirrored and rotated accordingly. In the course of the book, you will learn that galaxy rotation curves are among the central key observations for the phenomenon of dark matter and how laborious the search for a satisfactory answer to the question of the nature of this matter is. I hope you find this search as exciting as I do.

Telfs
June 2017

Wolfgang Kapferer

Contents

1
Introduction

If you search for answers to the question "What is dark matter?" in *the* information source of our time, the Internet, you will be well served quantitatively. The simple question **"What is dark matter?"** entered into a common Internet search engine will bombard you with about 20 million hits in a flash. I admit not to have read all pages found, but after a short review of the list of results, one must assume that dark matter is most probably something

- invisible,
- mysterious,
- dominant,
- to be believed in,
- …

These are all attributes that are difficult to combine with the methods of natural science. It quickly becomes clear that almost all large structures in the universe, such as galaxies, groups of galaxies and clusters of galaxies, predominantly consist of dark matter. And one learns in several treatises that no dark matter has yet been found directly on Earth or by space probes within our solar system. One reads that it must be a form of matter that hardly resembles the matter we are familiar with. It is so weakly interacting that it permeates every material in our world, almost without any sign of interaction with the elements we know. The adverb *almost* in the last subordinate clause is of particular importance. Because in order to detect dark matter directly in experiments, we need some kind of interaction with the particle physicists'

W. Kapferer, *The Mystery of Dark Matter*, Astronomers' Universe,
https://doi.org/10.1007/978-3-662-62202-5_1

instruments, apparatus from the type of matter we are familiar with. At least this is the hope scientists currently foster with their experiments.

The most successful theory of structure formation in the universe to date states that all the matter in our universe consists of about 85 % of this mysterious substance - dark matter. And it should not be unmentioned at this point that there is an even more significant player on the cosmological stage: dark energy, which according to the present knwoledge dominates the universe even more on large scales. However, as this book deals with the phenomenon of dark matter, dark energy will only be touched upon briefly.

A somewhat more intensive research on the Internet soon reveals to those interested that the term dark matter had already found its way into the literature of astrophysics in the 1930s and 1940s. If one takes the trouble to search for the number of all publications in which the term *dark matter* occurs in the text by means of a search engine for astrophysical literature (*NASA's Astrophysics Data System Bibliographic Services*), one gets as a result a search list with the "astronomical" number of almost 100,000 publications. About 67,000 of these are peer-reviewed publications (Fig. 1.1). But even more interesting than the sheer number of publications is the explosive

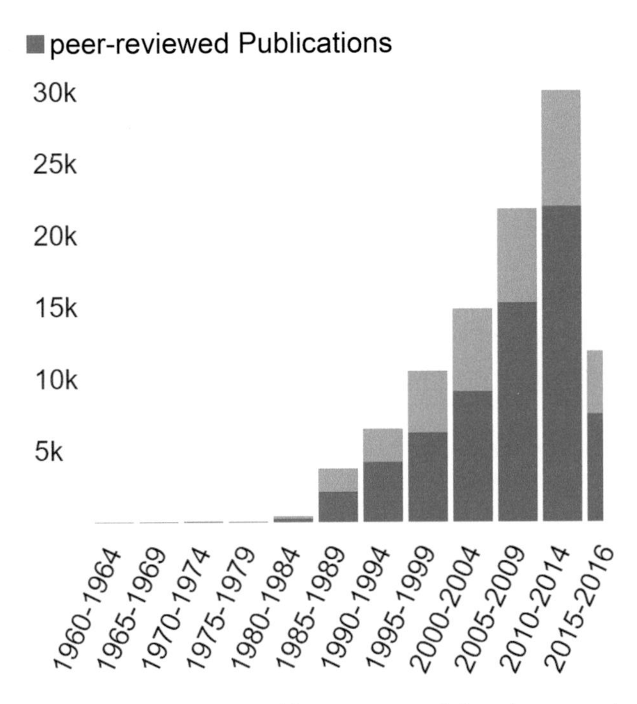

Fig. 1.1 Number of peer-reviewed publications containing the term *dark matter*. In the period 1960 to mid-2016, represented in units of 1,000. Source: *NASA's Astrophysics Data System Bibliographic Services*

increase in publications on dark matter starting in the early 1980s. To illustrate: In the years between 1930 and 1979, there were *in total only 428* publications in which the term dark matter appears.

Unfortunately, despite enormous efforts, the goal - the clear and direct detection of dark matter - has not yet been achieved. Not a single lump of this mysterious matter can be found in any astrophysical exhibition. And it seems that since the advent of the term dark matter in the astrophysical literature, the number of ideas and approaches to solving this problem has increased as rapidly as the presumed amount of dark matter in the universe.

Perhaps it is the same with the dark matter as once was with *ether* in the late 17th century. It was postulated in the models of these days as a medium for the propagation of light and gravity and was subsequently sought in vain by physicists for over two centuries. Nothing less than the successful theories of Einstein, Maxwell and Schrödinger mark the end of this long odyssey, describing the processes in our world completely without *ether*.

The aim of this book is to give the reader an understanding of the phenomenon of dark matter. It introduces the basic models of astrophysics, which lead to the conclusion that dark matter is the dominant matter-component in the universe. The book tries to answer the following questions:

- What major developments in the field of mass determination of astrophysical objects preceded the dark matter era? And above all, how is mass actually determined in astronomy?
- Which observations in the interplay to which models lead to the idea of dark matter?
- Which solutions without dark matter existed and why were they rejected?
- Which approaches and methods are currently being used to search for dark matter?
- Are there alternatives to the concept of dark matter?

A special focus in this book is on the methods applied in astrophysics. During my research, I found it amazing to see how many popular science books deal with topics like structure formation in the universe, speculations about parallel universes or the complex models of the first seconds after the Big Bang in only a few lines. Without a sound physical background, the danger of getting transfigured images of modern physics and astrophysics is very present. This must be prevented because what Marie von Ebner-Eschenbach (Austrian writer 1830–1916) once wrote is still relevant today:

Those who know nothing must believe everything.

2

The Art of Weighing a Star

What causes astrophysicists to postulate more matter than is observable? In order to answer this question, it is first necessary to digress into the history of natural sciences, explaining the concept of mass as used in the models of astrophysics and physics, in order to better understand the concept of dark matter.

The term *dark matter* already indicates what it is all about: matter that is not visible to the naked eye or to the sensitive instruments of astronomers. Matter that is literally "dark."

Especially when you consider that not too long ago only the range of light visible to the human eye was accessible to celestial explorers, it is not surprising that objects in the universe were still in darkness. However, even since we have been able to access ever larger areas of the spectrum of light as a source of information, the situation has not improved, only worsened.

The question that naturally arises at this point is: What causes astrophysicists to postulate more matter than can be observed? In order to answer this question, it is first necessary to make an excursion into the history of the natural sciences, explaining the term *mass* as it is used in the models of astrophysics and physics, to better understand the concept of dark matter.

W. Kapferer, *The Mystery of Dark Matter*, Astronomers' Universe,
https://doi.org/10.1007/978-3-662-62202-5_2

2.1 Mass Determination in the Cosmic Front Yard - The Solar System

The concept of mass as a measure of the gravitational attraction of bodies first appeared in astrophysics: in the exploration and mathematical modelling of the orbits of planets in the solar system. This is an example of how the astrophysical experiment, the observation, in combination with theoretical modelling leads to theories that enables us to predict the future state of astronomical systems. In the particular case about the orbits of the moons and planets of our solar system.

This exciting story began in the heyday of the Renaissance. Art and culture underwent a revolution and brought new ideas into the world, which would reverberate for centuries. During this time, our knowledge about the processes in the cosmos was realigned. The combination of detailed observations and their evaluation using theoretical-mathematical models made it possible to calculate the planetary orbits, predict their positions on the firmament and thus transfer them from the world of myths to the scientifically rational world of the Enlightenment. And, as almost always in the history of science, it was not a simple, direct path to the first useful models of planetary motion; it was rather a tortuous path of trial and error, as complicated as the planetary orbits in the night sky itself.

In our history, three great natural scientists and their work are in the focus of our interest: *Tycho Brahe, Johannes Kepler* and *Sir Isaac Newton*. Of course, this is an abbreviated approach and any historian of natural sciences would have to dismiss this approach as inadequate, but for the introduction of the concept of mass it seems sufficient to me at this point.

Let us first approach the person Tycho Brahe. He was one of the most important astronomers of his epoch. He studied philosophy, rhetoric, law, humanities and natural sciences from the middle of the sixteenth century at the universities of Stockholm, Leipzig, Wittenberg, Basel and Rostock - a true universal scholar of his time. The fact that he was able to represent his points of view vehemently was already shown in a duel at the age of 20 years, in which he lost part of his nose. According to many sources, it was a mathematical problem that he was truly passionate about, literally fighting out with his fellow students. However, he did not go down in the annals of the history of science because of his special dueling skills, rather he was one of the greatest astronomers of his time because of his exact observations of the sky. He made these observations completely without the astronomical telescope of Galileo Galilei, but only with - from today's point of view - primitive technical means, such as the wall quadrant (see Fig. 2.1). This

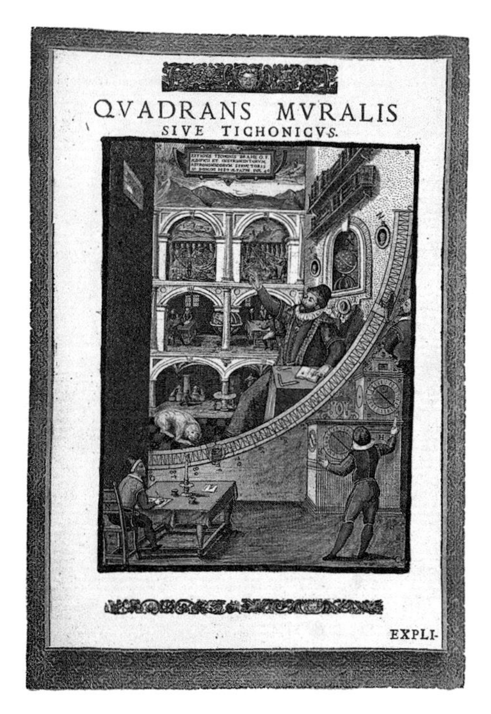

Fig. 2.1 Tycho's wall quadrant

is a simple but very effective instrument for measuring the positions of the stars on the firmament. The wall quadrant provides accurate viewing angles on a north–south axis. This allows the measurement of the highest position of an object on its path across the sky and the time of this event. These are two important quantities to describe the orbit of a celestial body in the sky. If you like, it was a large protractor for celestial objects.

Exact measurements of the courses of the wandering stars, as planets were called at that time, were scarce at that time. Tycho Brahe, with the help of several employees, collected data on the positions of celestical bodies on the firmament in a quality and quantity as never before. Brahe wanted to refute the burgeoning heliocentric world view at his time with his excellent data. He rejected the idea that the Sun was at the center of the universe and that the earth rotated not only around the central star but also around its own axis. He said that this could not be reconciled with his everyday observations either. That's why he developed his very own world view, the *Tychonian Planetary Model*. It described the orbits of the planets and the Sun in such a way that the earth remained in the center of events and the Sun and planets revolved around it. In order to be able to reconcile all his observations with this model, he had the planets known at that time (Mercury, Venus, Mars, Jupiter and Saturn) all

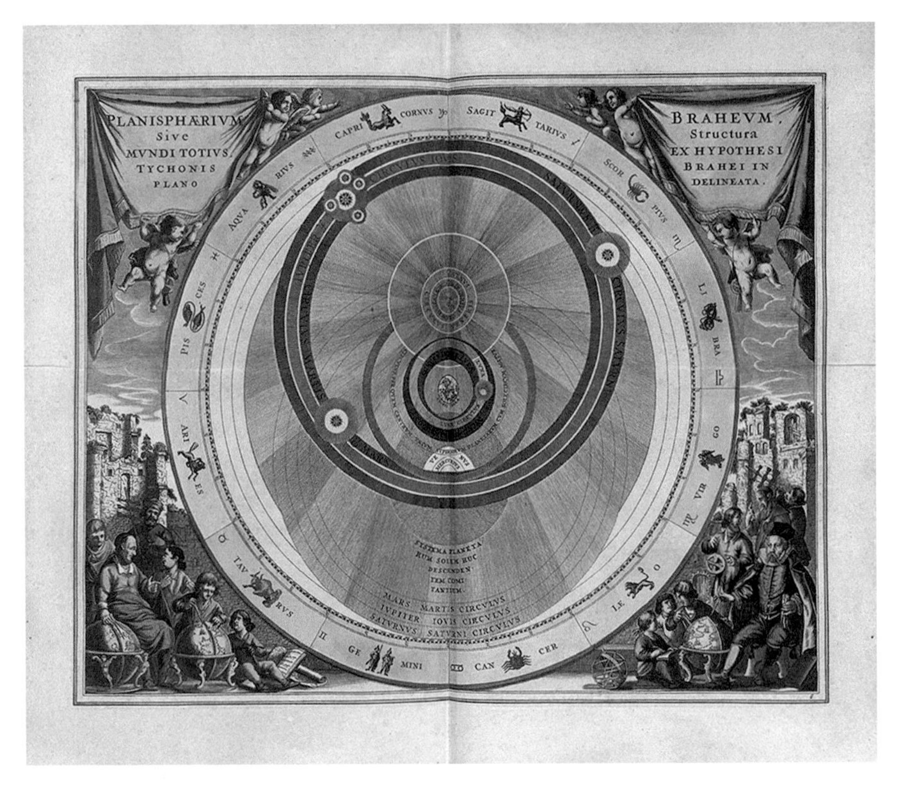

Fig. 2.2 The Tychonian Planetary Model. Source: Andreas Cellarius: Harmonia macrocosmica seu atlas universalis et novus, totius universi creati cosmographiam generalem, et novam exhibens. 1661

revolve around the Sun, see Fig. 2.2. The model was very complex and mathematically hardly controllable. A myriad of parameters were required, which almost always indicates that the model does not describe reality well.

And this brings us to the second protagonist of our story. So Brahe was missing a stable, elegant, theoretical *construction* based on the mathematics of those days, to help his model achieve a breakthrough. Therefore he hired Johannes Kepler as an assistant. Kepler was known and respected for his mathematical abilities. But the cooperation was not fruitful in the sense of Brahe. Kepler was not a good observer and, what was even worse, he was not convinced of the world view of his financier. And Brahe was afraid that Kepler might achieve fame on the basis of his great observations and that he would not receive the appropriate recognition.

Kepler, on the other hand, a number mystic, saw primarily mathematical relations as the underlying order of nature. Thus he initially thought that the planetary orbits followed five perfectly nested spheres within regular

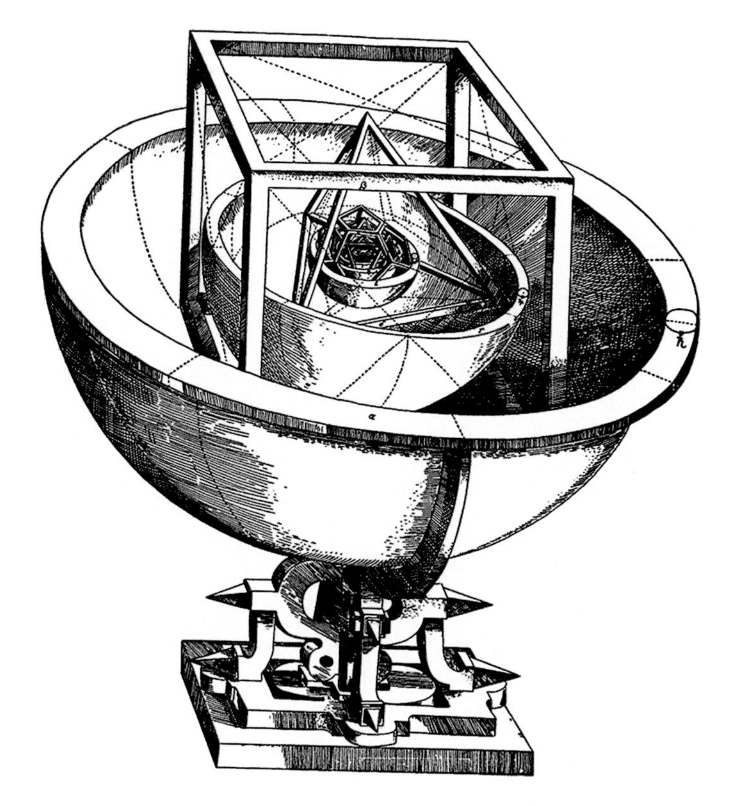

Fig. 2.3 Kepler's initial model of the solar system. *Mysterium Cosmographicum (1596)*

polyhedra (see Fig. 2.3). Different geometric shapes should describe the orbits of the planets exactly.

For him, as for many scientists of that time, geometric forms stood for a divine harmony, which had to be reflected in the paths of the stars. Today, one can hardly imagine the inner conflicts Kepler must have had when he found something other than this pure harmony in all the wonderful data of Tycho Brahe. Namely those relationships that we know today as the three Keplerian laws.

It was Kepler who recognized three fundamental connections in Brahe's data (see Fig. 2.4).

- The planets move on elliptical orbits with the Sun at one focal point.
- The imaginary direct connecting line Sun to planet sweeps over equal areas at equal times.
- The squares of the orbital periods of two planets behave to each other like the cubes of their large orbital axes.

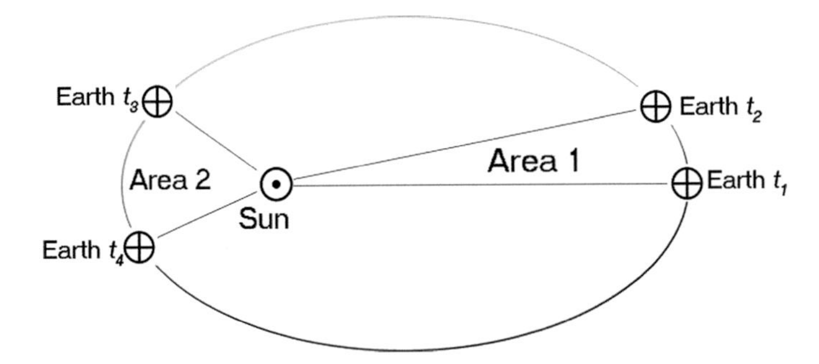

Fig. 2.4 Elliptical orbit of the earth around the Sun (strongly overdrawn, in reality almost circular) with our central star in one of the two elliptical foci. The time that the earth takes between the points t_1 and t_2 respectively t_3 and t_4 is equal. According to Kepler's second law - the imaginary connecting line between the Sun and the planets sweeps over equal areas at equal times - area 1 and area 2 are thus equal in size

But Kepler was not yet satisfied with these groundbreaking findings. He tried to find a causal effect for his observations. The natural scientist Kepler assumed that the Sun had a kind of magnetic effect on the planets. He himself described this effect as *anima motrix*, as *soul of the mover*, and already modelled for this long-distance effect a dependence entirely in the manner of the decrease of the intensity of the light with the distance from the shining star, see Fig. 2.5. This new force had been postulated by Kepler as the causal effect of the planetary orbits, which proved to be a very successful concept.

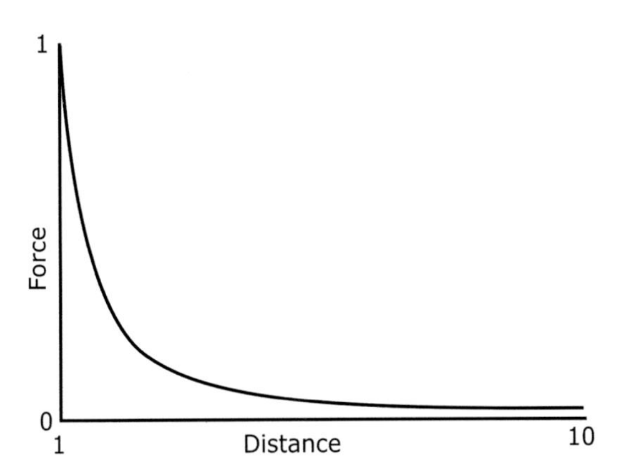

Fig. 2.5 Anima motrix distance behavior $1/r^2$. Assume that the attracting force has the value 1 at distance 1, and that it decreases quadratically along the drawn curve as the distance increases. At 10 times the distance the force is only a hundredth of the original value, but this force never disappears completely

Furthermore, in his model this force emanated from all bodies: the Sun, the planets, the moons, every lump of matter, no matter how small. Such a "magnetic" force effect of the bodies, the anima motrix, with its square decrease, has another interesting property: it will never completely extinguish. Even in the deepest depths of space, the Sun has its effect according to this model, but of course only to a very small extent. Yet this force will never disappear completely.

In *Astronomia nova,* 1609, Kepler's laws were not yet elegantly mathematically modelled, but rather formulated as loose axioms - a circumstance that was to change at the end of the seventeenth century.

With Sir Isaac Newton, the third actor in our history, and his major work, the *Philosophies Naturalis Principia Mathematica* (1686), a theory of the dynamics of bodies based on mathematical description appears for the first time at the stage of natural sciences. His theory formally brought together the observations of Galileo Galilei's fall-experiments and Kepler's laws of our nearest celestial bodies, thus creating a generally valid theory of gravitation (mass attraction). Newton's theory was unchallenged for several centuries and in many areas it still is today. It can be used to describe a large number of observations simply and elegantly.

Again, there are three connections, three laws, which describe the movements of the bodies and allow exact predictions of their paths:

- **Newton's first law or principle of inertia:** A body maintains its speed and direction as long as it is not forced to change its state of movement by external forces.
- **Newton's second law or principle of action:** The change in motion is proportional to the action of the force causing it and occurs in the direction of the straight line along which that force acts.
- **Newton's third law or interaction principle:** Forces occur in pairs. If one body exerts a force on another body (actio), a force of the same magnitude but in the opposite direction (reactio) - actio est reactio - acts in reverse.

The first important finding in Newton's model: The acceleration of bodies is caused by a force. In this abstract concept, a force always causes a change in direction and speed of the body's path.

The stronger the force, the stronger the acceleration, the stronger the change of path. Newton's second law describes the change of motion *proportional* to the acting force. So in equal proportion to a quantity that we could call **Mass**. This description also fits well with our everyday experiences. Just think how far you can throw a tennis ball or a medicine ball with the same force. According to Newton's first law, every application of force is accompanied by a change in speed or direction of a body, i.e. an acceleration, so mass can be understood as the ratio of force to the resulting acceleration.

Let us note: Already at the end of the seventeenth century there was a general theory of the movement of bodies, whose central concept is an assumed force, gravity, which causes orbital changes. Every body emits this force and every body is influenced by this force. The resulting forces accelerate bodies, which we can determine by measuring their positions in the sky at different times. The strength of the acting force emanating from a body and how strongly it reacts to an external force depends on its mass. Together with the principle of superposition - the sum of the individual effects results in an overall effect - the natural scientists were thus given a set of rules that allowed them to make exact predictions about the movements of the celestial bodies in the cosmos known at that time.

In order to be able to apply this set of rules to our solar system, a further insight is needed. The realization that an orbit like that of the planets around the Sun is an accelerated orbit at all times. It is decisively influenced by our Sun, with its large mass and, consequently, its equally large gravitational effect on its surroundings. It is this force that literally forces the planets to follow the elliptical orbits around the Sun that Kepler recognized.

Actually, at this point one should think that all planets should simply fall into the Sun because of its enormous gravitational pull, and that therefore there should not actually be any planets. However, nature is - fortunately - generally more complex and at this point the history of the origin of our solar system should not be left completely out of consideration.

After the Sun and its planets formed from the gravitational collapse of an interstellar gas cloud, the resulting system had an angular momentum that caused the objects to rotate uniformly around their common center of mass (roughly the position of our Sun). In the case of planetary orbits this means: The closer the object is to the central star with the dominating mass, the stronger the attraction of the central star is, but at the same time the higher is the orbital speed of the object, which counteracts this attractive acceleration, see Fig. 2.6.

At this point, an interesting thought-experiment is very helpful: If the Sun were a "dark" star not directly observable, we could draw conclusions about this dark mass and its position in our solar system solely from the changes in speed along the planetary orbits. What a powerful concept to bring light into the darkness of our universe.

Let us recap: Bodies are themselves sources of an attractive force, gravity, and in turn react to the gravitational force of other bodies. All in all, the

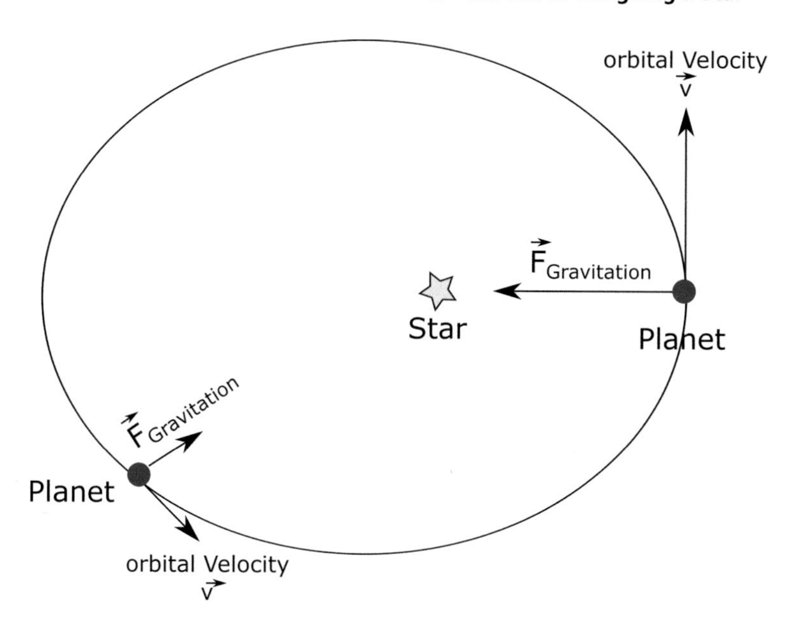

Fig. 2.6 The force acting on a planet moving in a stable orbit around its central star is determined primarily by the mass and distance to the central star: the closer, the greater the gravity. The orbital velocity $\vec{v}$ is a function of the distance of the planet from the central star: the closer to the central star, the higher the orbital velocity. These two opposing causes force the planets to move in elliptical orbits

sum of all acting forces causes a resulting acceleration, which we can measure by observing the celestial bodies at different points in time. The concept of mass, which describes acceleration in relation to force, is the central anchor point for accurate predictions of the stars' orbits. If we can measure the masses and current positions of the celestial bodies, then we can accurately predict their future positions. This capability enables us to send not only artificial satellites for telecommunications, earth exploration and astronomical observations, but even manned space capsules into the nearby cosmos.

2.2 Mass Determinations in the Galaxy

Equipped with the tools of Newton and Kepler, we can leave the solar system and ask ourselves the question: How large are the masses of the stars in our galactic environment?

If we want to apply Kepler's method, we need objects that move on measurable trajectories. Observations of the stars in our galactic neighbourhood have shown since the end of the eighteenth century that stars not only occur

Fig. 2.7 A panoramic view of our Milky Way. Picture credits: ESO/S. Brunier

isolated like our Sun, i.e. far away from neighbouring stars, but are often in binary or multiple systems. This proximity in multiple systems results in high accelerations. And you guessed it, you can determine them by measuring the positions of the stars on the firmament at different times. The first binary star catalogue was published as early as 1779 by the court astronomer of Mannheim, Christian Mayer, in his text "De novis in coelo sidereo phaenomenis in miris stellarum fixarum comitibus," in which he mapped 72 binary stars. From today's perspective, supported by star formation models, it is assumed that about 50% of all stars are members of binary or multiple systems. So when we observe our Milky Way, see Fig. 2.7, in the night sky, every second star is in a multiple system. This fact helps us immensely in determining the mass of the stars. For, if one can measure the changes in the positions of the celestial bodies relative to each other, one can use the tools of Newtonian mechanics to deduce the masses involved.

There are different types of binary stars. One species, however, plays a historically special role on the stage of mass determination of stars because of its easier identification: the visual binary stars.

Visual Binary Stars

This group of binary stars describes those that can be observed optically separated with a telescope. We will first take a look at a well known and well researched binary star system: Sirius A and Sirius B. Sirius A, also called Dog Star, is the brightest star in the night sky and can be found at a distance of about 8.6 Ly (light years) from us in the constellation Big Dog. That Sirius is a binary star system was first noticed in 1844 by Friedrich Bessel, a nineteenth century astronomer and mathematician. Bessel recognized a strange proper motion of the celestial body of Sirius A in the night sky in the long-time position data. As a possible reason for these changes in position, Bessel assumed a companion star, an unobservable mass in its vicinity, i.e. dark matter from the point of view of the time. Bessel believed that if there were visible stars, nothing spoke against the existence of invisible ones. It was not until about twenty years later, in 1862, that the US astronomer and telescope builder Alvan Graham Clark was able to optically resolve the binary star system Sirius. For this discovery he was awarded the Lalande Prize of the French Academy of Sciences in the same year. Today, of course, binary star systems can be studied with space-based telescopes such as the Hubble Space Telescope in much better resolution (see Fig. 2.8), but the physical models we apply to our data today are the same as in the days of Bessel or Clark. With the help of the observations, the orbital parameters of the binary star

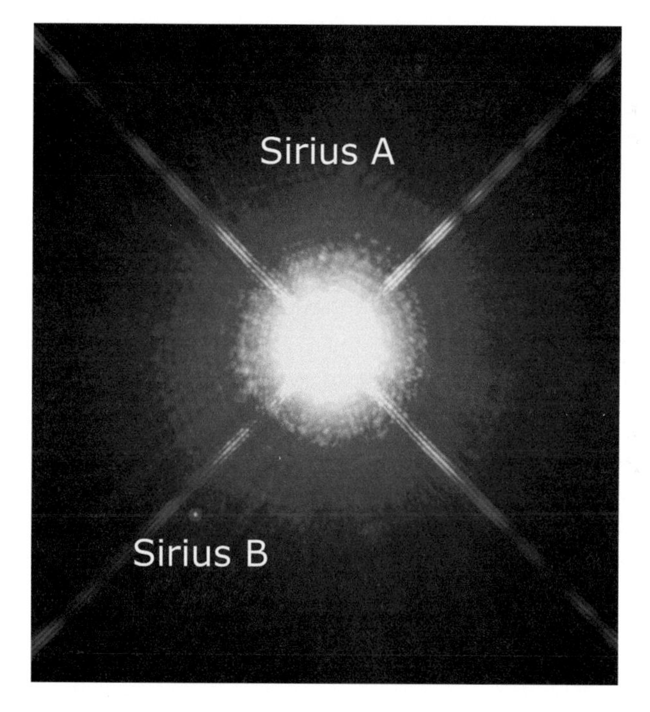

Fig. 2.8 Sirius A and B, observed with the Hubble Space Telescope. Picture credits: NASA, H. E. Bond and E. Nelan (Space Telescope Science Institute, Baltimore, Md.); M. Barstow and M. Burleigh (University of Leicester, U.K.); and J. B. Holberg (University of Arizona)

system determined from them and the principles of Newtonian mechanics, it is possible to determine the masses involved.

We will now look at this in detail for the Sirius binary star system. According to Kepler's third law, the squares of the orbital periods of two planets are equal to the cubes of the major orbital axes. It was none other than Newton who showed that Kepler's third law was a special case. Namely, the case where the mass of the Sun far exceeds the mass of all the planets taken together. In our solar system, this ratio is approximately 750:1, and if the masses involved deviate less strongly from each other, differently designed orbits can be observed.

In general, the orbits proceed around the common center of mass of all objects involved. In the case of our solar system, the center of mass is more or less at the position of the Sun. In the case of the binary stars, a different picture emerges. For the Sirius system, the center of mass between the two stars is located at point M, as shown in Fig. 2.9.

If the observed positions are translated in such a way that the more massive Sirius A is fixed in the sky and the lighter Sirius B moves relative to it, one obtains the so-called *relative true path* of the overall system. This so-called coordinate transformation allows us to describe the system by means of a virtual Kepler orbit.

If the orbital parameters have now been determined by many observations, the total mass of the system can be derived. But the attentive reader will notice at this point that we can determine the great semi-axis absolutely only if we know the distance of the system to us. Here we still have to work on one more step: the distance measurement by means of the star parallax. It is the most important method to determine the distance of the stars in our environment.

Figure 2.10 sketches the basic idea of this so important distance determination. One observes a star which changes its position relative to other, fixed, stars in the course of half a year. Here star 3. The observed positions are shown in Fig. 2.10 in the boxes at time 1 and time 2. Since we have the distance a of the Earth from the Sun, it is now possible to calculate the distance to star 3 using trigonometry - surveying and mapping on somewhat larger scales.

The distance to star 3 in our example would then be $r = a/\tan\alpha$ with r the distance between our Sun and the star, a the distance between the earth and the Sun, and α, the measured angle relative to the distant fixed stars (star 1 and 2). So much for the theory. The problem lies in the angular resolution of the astronomical observations, which is illustrated by a numerical example of our binary star system Sirius. We now know that the system is located at a

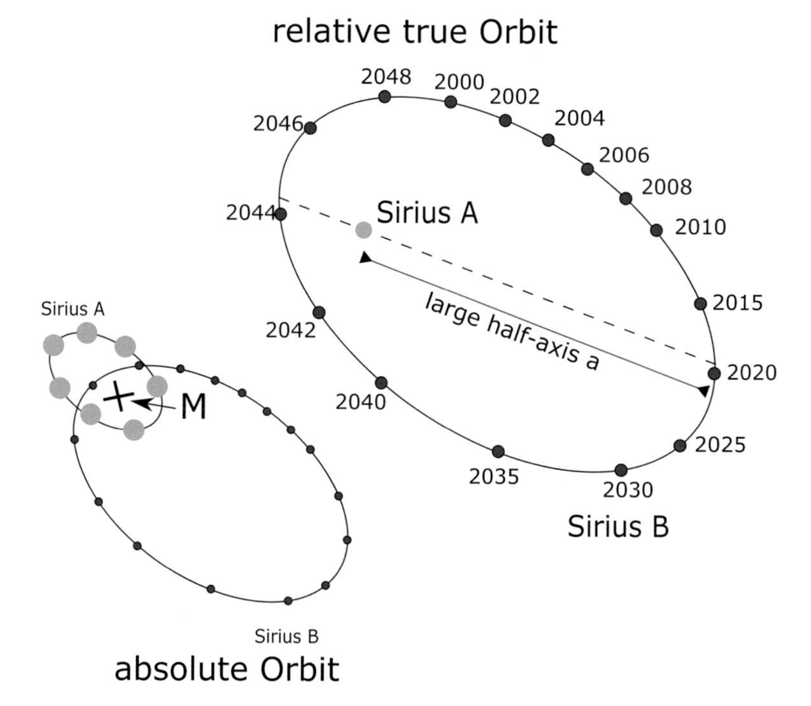

Fig. 2.9 The orbit of Sirius A and B with an orbital period of almost exactly 50 years. Lower left the observed absolute orbit of the two stars on the firmament. Upper right is the relative true orbit of Sirius B around Sirius A. The dotted line connects the most distant orbital point (apastron) of Sirius B with the point of least distance (peri-astron) from the fictitious central star. The large half-axis a is shown in red

distance of about 8.6 Ly from us. Calculated in kilometers, this is about 8.6 times 9,460,000,000,000 km, which is about 80,000 billion kilometers. The average distance of the Sun from the earth corresponds to a *astronomical unit* which is about 150,000,000 km. Thus the angle α is only about 335 milli arcsecond, not even half a second of arc. If one were to measure a parallax of exactly one arcsecond, the measured star would be at a distance of 3.26 Ly, this distance is called *parsec* - from parallax second - and marked with the abbreviation *pc*. It is the common unit for distances in the depths of space.

And again it was Friedrich Wilhelm Bessel who was the first to make this observation. First it was not Sirius but a striking "fast-moving" star called 61 Cygni, which he was able to determine with this method in 1838 at a distance of about 10 Ly. At present, satellite telescopes such as Hipparcos (1989–1993) and Gaia (launched in 2013) can be used to make exact parallax measurements on a billion stars in our Milky Way and in some cases even to determine parallaxes in the vicinity of our galaxy.

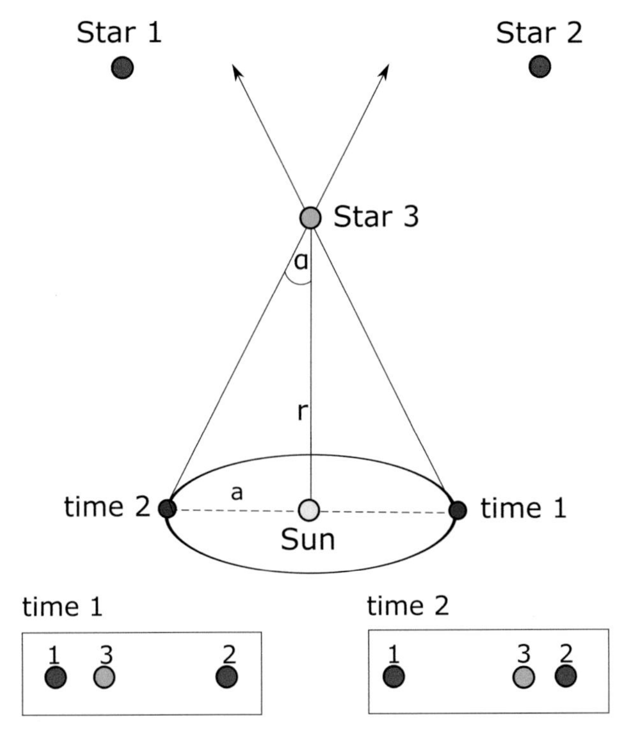

Fig. 2.10 Star parallax: A foreground star (star 3) is observed at two distinct points in time. Between these points in time the earth moves around the Sun, thus changing the projected position of star 3 to the far away background stars 1 and 2 (point in time 1 and point in time 2). From this observation the distance to star 3 can be determined by trigonometry

But back to our binary star system Sirius, from which we want to determine the mass. To determine the total mass, we have everything necessary at hand and nothing more stands in the way of an evaluation. The result for the entire system is about 3.2 solar masses. Further knowledge of the ratios of the large semi-axes in the absolute orbit of the two stars around their center s of mass allows the ratio of the masses to be determined: $\frac{a_A}{a_B} = \frac{m_B}{m_A}$. For the system Sirius, it results in masses of 2.3 solar masses (Sirius A) and 0.9 solar masses (Sirius B).

Armed with Newtonian mechanics, excellent observations and solid trigonometry, we are able to determine the masses of binary stars in our galactic neighbourhood. It is interesting to note that already at the time of the first systematic star observations the existence of nonvisible companion stars was assumed to explain the relative motion of stars. If you like, they were "dark" companion stars that were postulated to explain the observed motions of some stars. These postulated companions were then directly confirmed in their existence by ever better observation instruments. Thus astrophysicists have - in this respect - historically a very positive attitude about the postulate of dark masses. It is also possible to determine the mass of binary star systems, which can no longer be resolved visually one by one, by examining the total light of the star system and studying periodic variations in brightness due to mutual occultations and inferring the corresponding orbital parameters. This type of binary stars are called eclipsing binary. Another group are spectroscopic binary star systems, which reveal their existence in the spectroscopic investigation of the emitted light. For the moment, however, we will be content with the Sun and visual binary stars; for this class of objects we have methods in hand to determine their masses. But how do we obtain the masses of all the other stars in our Milky Way that we cannot measure in this way? In order to answer these questions, we need an additional tool so that we can place the further discussion of the dark matter phenomenon on a solid foundation. The name of this tool is *mass–luminosity relationship*, a fundamental relationship in astrophysics.

2.3 Your Type Reveals Your Weight

The Mass–luminosity Relationship

The classification of the apparent brightness of stars is an ancient art of astronomy and goes back to Hipparchus of Nicaea (190 to 120 BC), a Greek astronomer. At that time the Greeks divided the stars, observed with the naked eye, into six different classes according to their brightness. The brightest stars were assigned to class one and the faintest stars to class six. In this classification it must be taken into account that a sensory stimulus does not increase linearly with the physical quantity, but only according to a logarithmic scale. This relation entered the literature as Weber-Fechner law in the middle of the nineteenth century, see Fig. 2.11. In short, a class one star observed with objective measuring equipment is not twice as bright as a class two star, but 2.5 times brighter, and already 6.3 times brighter than a class three star.

Since with the advent of telescopes many stars that had not been visible until then became observable, it was time to expand the scale accordingly. In the middle of the nineteenth century the brightness scale was realigned. The aim was to preserve the classification of the Greeks as much as possible. A class one star was now objectively, physically measured 100 times brighter

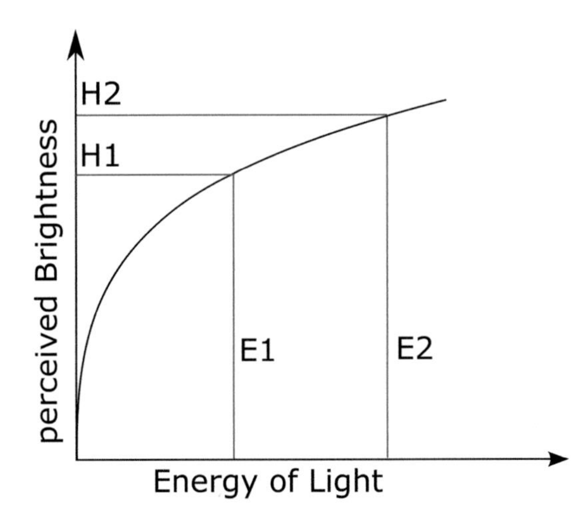

Fig. 2.11 Weber-Fechner law: Doubled energy of light is not perceived as twice as bright, but only as brighter according to a logarithmic scale

than a class six star. The scale was calibrated on the standard stars, which were characterized by a constant brightness over many years. Nowadays many of these precisely measured standard stars are catalogued. Of course, only the energy of the light of a star that reaches us can be measured. This energy decreases with the square of the distance, a circumstance that can be clearly illustrated purely geometrically, see Fig. 2.12. If, at a distance of 10 pc, about 32.6 Ly, from a star, the amount of light emitted is divided by an imaginary surface A and one carries out such a measurement again at double the distance, this area is covered by one quarter of the original imaginary surface A, leading to only 25% amount of light.

However, this relation can also be thought of in reverse. If we knew the amount of light from a star by an imaginary unit area at a unit distance (say 10 pc), we could derive the distance to this star by measuring its brightness here on Earth, a method known in the literature as the distance modulus. This magnitude is called absolute brightness in literature.

But now back to the masses of the stars - to the mass–luminosity relation. As described in the previous section, parallax measurement allows us to determine the distance to some of the binary star systems in our Galactic neighbourhood. Their apparent magnitudes and their distances are accessible by measurements with our telescopes, so the absolute magnitudes of the stars at a defined distance are known. Newtonian mechanics provides the mass of stars and thus the quantities for a mass–luminosity diagram. It is important to emphasize at this point that there are many technical

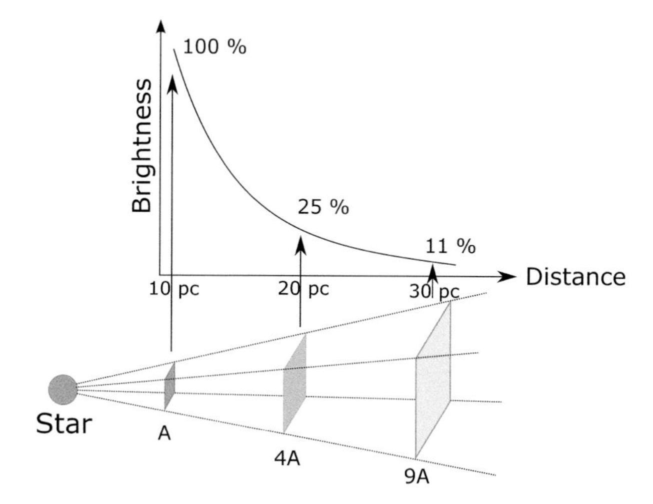

Fig. 2.12 Drop in brightness as a function of distance: The quadratic decrease in brightness is easy to understand when you consider that the amount of light emitted by the star illuminates a squarely growing area as the distance increases. In the example, the same amount of light falls on the area A at 10 pc distance as on the area 4A at 20 pc distance and on the area 9A at triple distance

calibration problems and that measurement error analyses are required - but this circumstance is not relevant for understanding the mass determination of stars.

In Fig. 2.13, the masses of some stars of visual binary stars (each for the more massive partner) are plotted against the absolute magnitudes (luminosities), the data are from the Hipparcos satellite mission.

At first sight we see that there is a strong correlation between mass and luminosity. This connection is very important, because it opens a window for us to determine the distance to all the other stars whose orbits we cannot measure. But we still lack a crucial ingredient to extend this relationship to all stars. We need an additional relation: As we will see, it is the knowledge of the spectral type of a star and the absolute magnitude that can be derived from it.

The Spectral Types of Stars

Up to now we have classified the stars according to their brightness, just like the ancient Greeks did, from the brightest stars in the night sky to the least luminous ones. But there is another classification that tells us much more about stellar objects and ultimately opens the door to a comprehensive

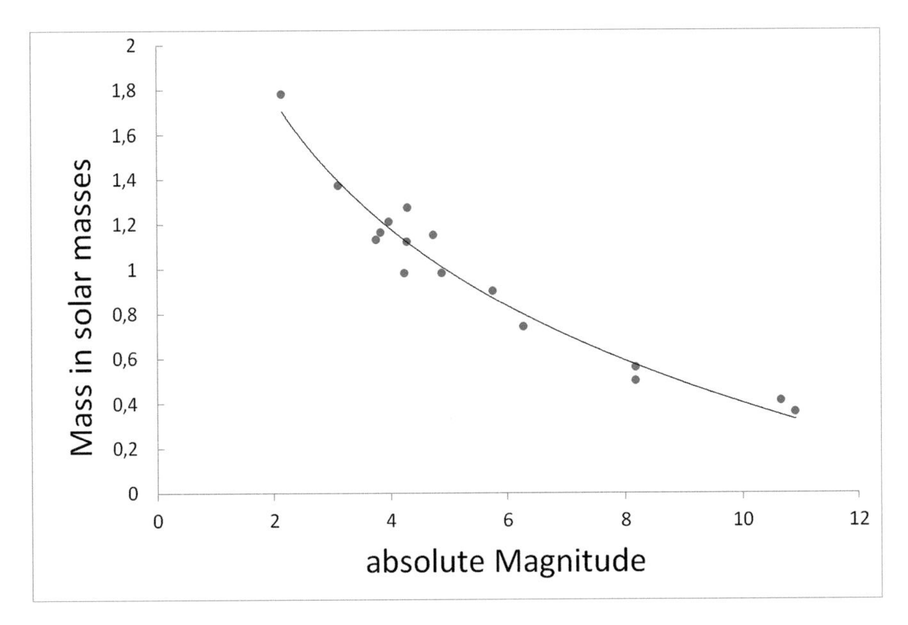

Fig. 2.13 Mass–luminosity relationship for the more massive main stars in binary star systems. The masses are given in solar masses, the absolute magnitudes (brightness) in 10 pc distance. Data from Soderhjelm (1999)

distance determination that allows us to reach out into the depths of space. It is the spectral analysis of the light of stars and the interesting relations that can be obtained from it. Again, an interplay of observation and theoretical model can be seen here, which in the course of the history of science has been able to revolutionize our limits of knowledge. However, for a better understanding we want to turn to the beginnings of the history of spectral analysis in astrophysics.

It began with the young orphan Joseph Fraunhofer (1787–1826), who lost his parents at the age of twelve. He was given a 6-year apprenticeship as a mirror grinder in Munich by his guardian and even survived his teacher's house collapsing. His salvation was observed by honorable people of those days who supported him from then on. After his apprenticeship he was able to join the Mathematical-Mechanical Institute of Reichenbach, Utzschneider and Liebherr in 1806. This institution was founded in 1802 for the production of astronomical and geodetic instruments. There Fraunhofer worked his way up to become an excellent optician. He was the first to systematically investigate the dark lines in the solar spectrum, but Fraunhofer was not the first to discover the spectrum of a light source. An Italian Jesuit, physicist and mathematician named Francesco Maria Grimaldi

(1618–1663) in Bologna already recognized the splitting of the colors of sunlight by the optical prism. Even Sir Isaac Newton had a try at a theory of light, based on a particle model, and it was Thomas Young (1773–1829) who finally formulated a theory of light based on waves.

But back to Fraunhofer, who brought a system into the description of the light spectrum of the Sun. Figure 2.14 shows some absorption lines in the solar spectrum and their categorization, the systematic description of the dark lines in the visible solar spectrum first introduced by Fraunhofer.

Before we turn to the dark lines in the solar spectrum, a very brief outline of spectroscopy and its methods is certainly helpful. The term spectroscopy was first coined in 1882 by Arthur Schuster (1851–1934) and comes from the Latin term *spectrum* for *appearance* and the Greek term *scopein* for *view*. So it is the doctrine of contemplating the apparition, which sounds quite pathetic. For the natural scientists at Sir Isaac Newton's time, however, it must have had an almost magical effect: the splitting of a white ray of Sunlight into its colour components by an optical prism. Figure 2.15 shows the spectrum of a white light beam through a prism. It is clearly visible that the light beam coming from above is refracted at the air-prism-air transitions and additionally split into the colour components. Additional reflections are visible and are caused by the geometric shape of the prism.

In order to understand the physics of this phenomenon, we will first look at the refraction of a light beam at the transition between two media. The model used to describe this phenomenon is based on the property of light that its propagation speed is different in different media. Let us assume that the speed of light propagation in a medium is c_1 and in an adjacent medium $c_2 < c_1$. If a light beam with the diameter D shines obliquely onto the interface, as shown in Fig. 2.16, then the left side of the light beam moves at the lower speed c_2 while the right side is still moving at a faster rate c_1 - the light beam is refracted.

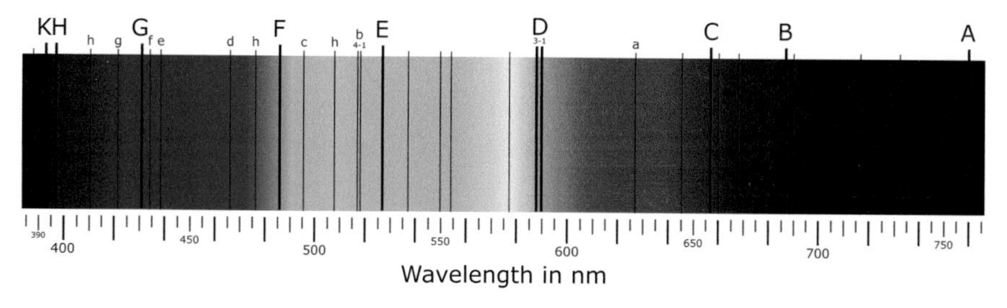

Fig. 2.14 Some Fraunhofer lines in the visible range of the electromagnetic spectrum of the Sun and their names

Fig. 2.15 Splitting of a light beam by a prism into its color components. Picture credits: By Kelvinsong (Own work) [CC0], via Wikimedia Commons

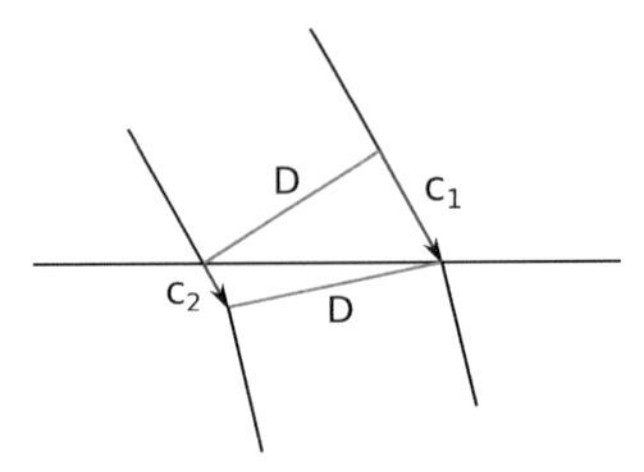

Fig. 2.16 If a light beam with the diameter *D* shines obliquely on a separating layer between two different media, with the propagation speed of light $c_2 < c_1$ in the media, then the light beam moves on the left side with the lower speed c_2 while on the right side he is still moving with the higher speed c_1 - the light beam is refracted

In a medium, the speed at which light propagates also depends on its wavelength, i.e. its colour. This is why there are different degrees of refraction for different wavelengths, blue is refracted more strongly than red. With this knowledge we can elegantly describe *appearances*, which Sir Isaac Newton has so admirably named *spectrum,* in a model.

When Fraunhofer examined the spectrum of sunlight more closely, he found several black lines in it, as if certain color components, wavelengths, were missing in the split light. The work of Gustav Robert Kirchhoff (1824–1887) and Robert Wilhelm Bunsen (1811–1899) made it possible to assign these gaps to chemical elements in the Sun and our atmosphere, in the literature these black lines are called absorption lines.

It is one of the great achievements of quantum mechanics to provide us with a model for the origin of these lines. Light of a certain energy can interact with elements such as hydrogen or helium. At certain energies, i.e. wavelengths of light, an atom is excited in such a way that the spatial configuration of the electrons changes. A light beam in this wavelength range gives off its energy to the electrons of the atomic shell and disappears at this point in the spectrum, *absorption lines* arise. However, this energy conversion can also work in the other direction: An excited atom spontaneously releases its excitation energy in the form of a quantum of light, one can see *emission lines* in the light spectrum here. Since the electron configurations of an element are not arbitrary, but only permit certain energy levels, the chemical composition of objects in the universe, but also of terrestrial matter, can be studied in detail in the laboratory by measuring the emission- or absorption lines. This is the basic principle of spectroscopy.

As can be seen in Fig. 2.14, Joseph von Fraunhofer observed several absorption lines in the solar spectrum. They come partly from the outer layers of the Sun and from gases in our earthly atmosphere. Fraunhofer gave these lines letters back then, but today we can assign these elements. A few are given in Table 2.1.

Like the lines in the spectrum of light, the energy distribution within the spectrum reveals physical properties of the emitting source. This is shown in Fig. 2.17.

It can be seen that the Sun has its maximum energy output in the green region of the spectrum at around 500 nm (nanometers - one billionth of a meter). The shape of this curve enables us to assign a temperature to the photosphere, i.e. the light-emitting layer of the Sun. The theoretical modelling of this curve is also one of the first major achievements in quantum mechanics by Max Planck (1858–1947) and is referred to in the literature as Planck's blackbody radiation. It is the energy distribution of radiation that a body emits only because of its temperature. In Fig. 2.17, it is drawn as a grey area.

But let us come back to the question of the mass determination of stars. With the help of Newtonian mechanics and the determination of orbital parameters of the stellar partners, we can determine the mass for a few binary stars in our galactic front garden. By parallax determination, the distance to

Table 2.1 Some Fraunhofer lines and corresponding elements

Fraunhofer line	Element
y	Oxygen
c	Iron
F	Hydrogen
d	Helium

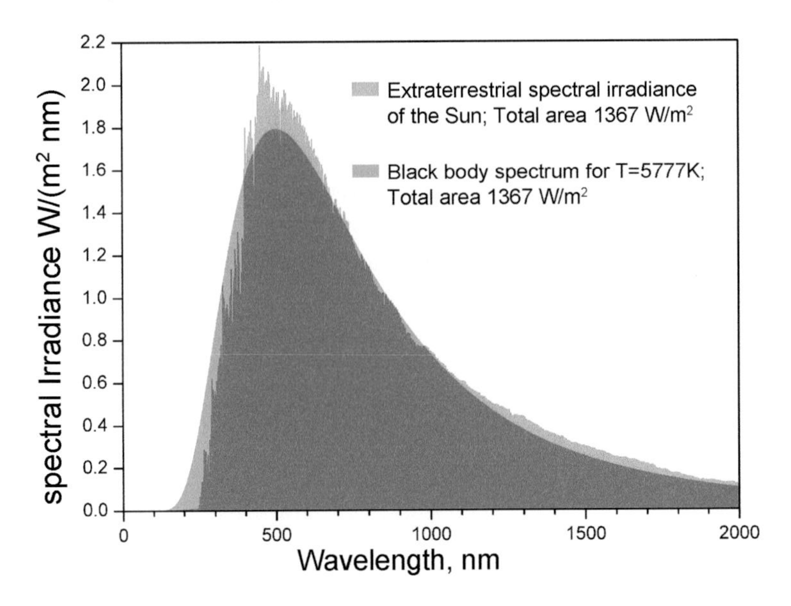

Fig. 2.17 The effective temperature of the Sun (5,777 K) is the temperature that a black body of the same size would have to have in order to emit the same amount of radiation. Source: By Sch [GFDL (https://www.gnu.org/copyleft/fdl.html) or CC-BY-SA-3.0 ()], via Wikimedia Commons

these binary star systems is known and thus their absolute brightness. The question now is: How do you get to the masses of all the other stars?

To do this, we need a physical model for stars, which must prove itself successfully on the observed binary star systems. The beginnings in the physical modelling of stars are marked by the work of Sir Arthur Stanley Eddington (1882–1944), a British astrophysicist. The model will be demonstrated on our central star. Our Sun has a mass of about 2×10^{30} kg. Its brightness has been constant since time immemorial, and its extent in the sky has not changed since the first records of mankind. We know from the spectrum that it consists mainly of the elements hydrogen (92.1%) and helium (7.8%) and has a surface temperature of about 5,500 °C. It is a giant, hot gas ball. Actually, the Sun should collapse due to its gravity, but it does not because of the opposing gas and radiation pressure caused by thermonuclear reactions inside. It is in a state of equilibrium. Radiation pressure, like gas pressure, can be described by the temperature of its source, whereas the gravitational pressure of a star towards the center can be described by its mass. But all this is in equilibrium for stable stars. If one mesaureand is known, one can derive the others. The following conclusions can therefore be drawn in very brief form: If you know the spectrum of a star, you have information about its chemical composition and its surface temperature, and you can derive its mass from the equilibrium calculus. Once the mass of a star has been determined in this way, its absolute brightness (the sum of its total radiated energy per unit time) can also be determined theoretically. The formulation of this connection was the achievement of Sir Arthur Stanley Eddington, see Fig. 2.18.

So the key for determine the mass of stars in a fast ways is to determine their absolute magnitudes. But how can we reliably determine the absolute magnitudes of stars? With the telescopes of astronomers only the apparent brightness can be measured directly.

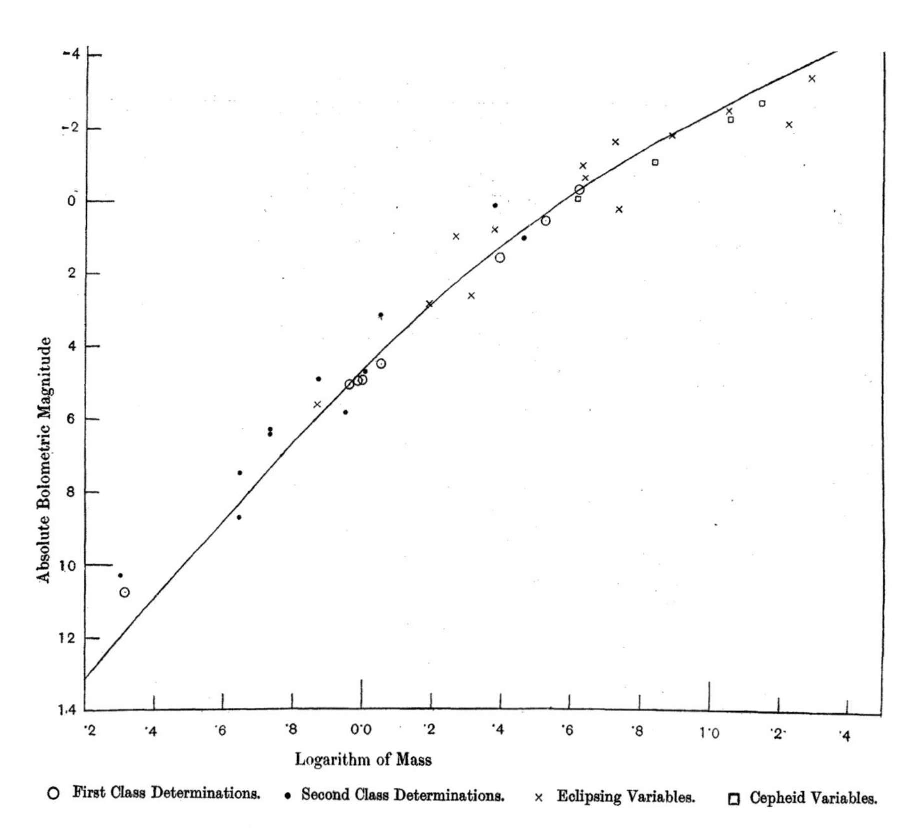

Fig. 2.18 Mass–luminosity relationship, original graphic by Arthur Stanley Eddington (1926). The *x*-axis is in logarithmic scale in units of the solar mass, the *y*-axis shows the absolute brightness. The plotted symbols correspond to the points of some stars already known, such as visual binary stars. The line is taken from the star model calculations of Sir Eddington

You guessed it, the key is in the spectrum of a star. Historically, the work of the Dutch astronomer Jacobus Cornelius Kapteyn (1851–1922) should be mentioned here as a pioneer. Kapteyn studied the proper motions of stars in the neighbourhood of our Sun and published a list of accurate parallax measurements. The list included some characteristics of the stars observed, including their absolute magnitudes and spectral types, a summary of certain properties in the stars' spectra. The classification is made into classes, common lines of elements in the spectra of the stars and the surface temperature of the stars were used as classification characteristics. In Table 2.2, some few properties of the spectra are listed and in Fig. 2.19 some corresponding spectra are shown. In addition, the surface temperatures of the spectral types are given, which corresponds to the maximum in the energy distribution of the spectrum.

Now, when the astronomer and chemist Ejnar Hertzsprung (1873–1967) and somewhat later the astronomer Henry Norris Russell (1877–1957)

Table 2.2 Stellar spectral types and their characteristics

Spectral type	Characteristics
O	Dominance of absorption lines of single ion helium
B	Helium lines become weaker
A	Hydrogen lines maximum
F	Hydrogen lines become weaker; Occurrence of iron lines
G	Iron lines and calcium lines
K	Titanium oxide lines and lines of simple molecules
M	Titanium oxide lines dominate

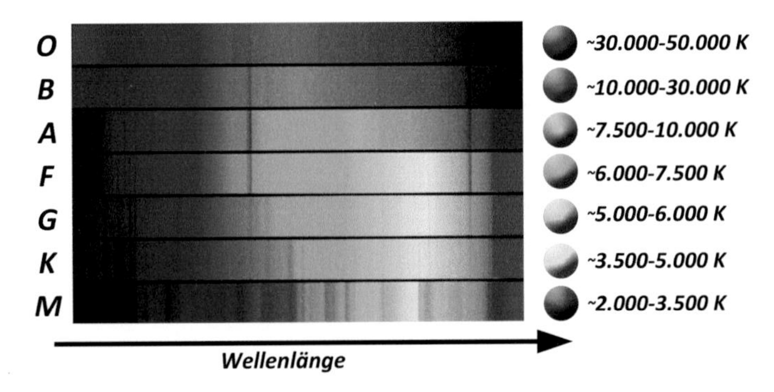

Fig. 2.19 The spectra of the spectral types - O, B, A, F, G, K, M - and their surface temperatures in Kelvin

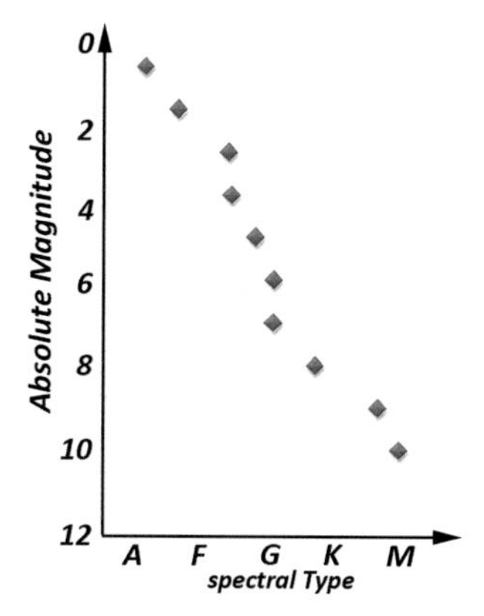

Fig. 2.20 Absolute magnitudes versus spectral types of some stars well measured by Kapteyn. Data from Russell (1914)

examined more closely the surface temperatures and spectral types and the absolute magnitudes of some well measured stars as determined by Kapteyn, an interesting and important correlation became apparent - the absolute magnitude and the spectral type are correlated. This connection becomes obvious in Fig. 2.20. The points correspond to the absolute brightnesses measured by Kapteyn, which are derived from parallax measurements and are therefore quite robust. One can see that stars with hotter surface temperatures, spectral type (A), are brighter than those with cooler surfaces, spectral type (M). In the astrophysical literature, this correlation is known as *Hertzsprung-Russell* diagram.

This relationship is fundamental to stellar astrophysics and allows us to elegantly determine the mass of most stars.

We can now roughly sketch the way to a mass determination based on the spectrum of a star. First you extract the spectrum from the light of a star. Its spectral class is determined by the distribution of the lines and the surface temperature. Via the *Hertzsprung-Russell* diagram, you then get its absolute brightness. If you have determined the absolute birthness, you can use the mass–luminosity relationship to infer the mass of the star. In short, this is the standard way to derive the mass of stars from their light.

Summary

The procedures outlined in this chapter for mass determination of planets and stars in our galaxy are exemplary in astrophysics. First, a model of mass is needed, essentially given by classical mechanics founded by Sir Isaac Newton, which was later so brilliantly embedded in a more general model by Albert Einstein (1879–1955) in his General Theory of Relativity. If we know the masses of the bodies involved and their directions of movement, we can theoretically calculate the locations of the objects at any future and past time, so mass is one of the keys in teaching about the movement of the bodies. Where we can measure motion directly, it is possible to determine the masses, for example in our solar system by the measurements of a Tycho Brahe and the evaluations of a Johannes Kepler, or for nearby binary star systems by determining the orbital parameters and a measured parallax. As technological progress made it possible to measure the first spectra of stars, a correlation was found between absolute magnitude and the spectral type of a star. Again, a model of the stellar structure was needed that described gravity and radiation pressure in a star's equilibrium state. This was achieved by Sir Arthur Stanley Eddington, to whom we owe the mass–luminosity diagram. The lines in the spectrum of stars, which were first precisely measured in the solar spectrum by Joseph von Fraunhofer, enable us to determine the spectral type of a star. In addition, the shape of a spectrum tells us the surface temperature of a star, an achievement of Max Planck's then still young quantum mechanics. This makes it possible to divide the stars into types according to their spectrum. Through the work of Ejnar Hertzsprung and Henry Norris Russell, a correlation between the spectral types of stars and their absolute luminosity was identified and this was the last piece of the puzzle to determine all the masses of stars in our galactic neighbourhood. The next chapter will show that something amazing came to light, you already guessed it - dark matter.

References

Eddington AS (1926) The internal constitution of stars

Russell HN (1914) Relations between the spectra and other characteristics of the stars. PopA 22:275–294

Söderhjelm S (1999) Visual binary orbits and masses POST HIPPARCOS A&A 341:121–140

3

The First Signs of a Big Problem

... It is a good rule not to put overmuch confidence in a theory until it has been confirmed by observation. I hope I shall not shock the experimental physicists too much if I add that it is also a good rule not to put overmuch confidence in the observational results that are put forward until they have been confirmed by theory.

Sir Arthur Stanley Eddington (1882–1944)

British astrophysicist

New Pathways In Science

Messenger Lectures

Cambridge at the

University Press, 1934

This quotation from Sir Arthur Stanley Eddington, a British astrophysicist who was the first to observe Einstein's theoretical model predictions of general relativity during a solar eclipse and who introduced the mass–luminosity relationship of stars, describes very aptly the manifestation of the dark matter phenomenon in modern astrophysics. The concept of dark objects was first introduced by Friedrich Bessel in the middle of the nineteenth century during his observations of the proper motion of a binary star. Bessel conjectured, encouraged by the teaching of the motions of bodies - Newtonian mechanics - that dark stars would probably have to exist in order to understand the strange motions of Sirius in the sky. Indeed, Bessel was right, with better observation-instruments some of these dark objects have been

W. Kapferer, *The Mystery of Dark Matter*, Astronomers' Universe,
https://doi.org/10.1007/978-3-662-62202-5_3

revealed as faint stars. The dark star/matter phenomenon of those days was more or less a problem of limited observations, which, from the point of view of that time, could in principle be solved by technological progress.

Precise measurement of astronomical objects and prediction of their movements using a theory of the dynamics of gravitationally interacting bodies is the task of celestial mechanics. In order to describe the observed orbits of the stars, a dark component, a dark partner, has often been introduced. These postulated objects were then mostly discovered by means of better observation techniques, a triumph for Newtonian mechanics. Observations that were not consistent with this theory were dismissed as shortcomings of the observation techniques. In this light, it is not surprising that the first signs of dark matter were considered a problem of astronomers' inadequate instruments for a long time.

Even in his early studies of the distribution of stars in our galactic environment, the British physicist Sir James Hopwood Jeans (1877–1946) came across the need for a considerable amount of invisible matter in order to reconcile the observations with Newton's celestial mechanics. In 1922, he published an article in the renowned scientific journal *Monthly Notices of the Royal Astronomical Society* in which he discussed the question of the motions of the stars in the then known universe. According to Jeans, the article was based on a masterly summary of the state of knowledge at that time about the distribution of stars in the universe by the Dutch astronomer Jacobus Cornelius Kapteyn. In order to put this work into context, it is important to remember that at that time, little was known about the structure of our Milky Way and the existence of other galaxies. The different interpretations of the data at the beginning of the twentieth century can be seen very well in *the great debate*, also called *Shapley–Curtis Debate*. This discourse between the astronomers Harlow Shapley (1885–1972) and Heber Curtis (1872–1942) was publicly held on April 26, 1920 at the National Museum of Natural History in Washington D.C. However, first let us take a brief look at the state of knowledge about the Milky Way at the time of Sir James Hopwood Jeans so that we can better understand his discovery and its significance in the discussion at that time.

3.1 Early Signs in the Milky Way

At the beginning of the twentieth century, the picture of our galaxy was very vague. The Dutch astronomer Cornelius Easton (1864–1929) published a paper at the turn of the century in which he showed that the Milky Way has

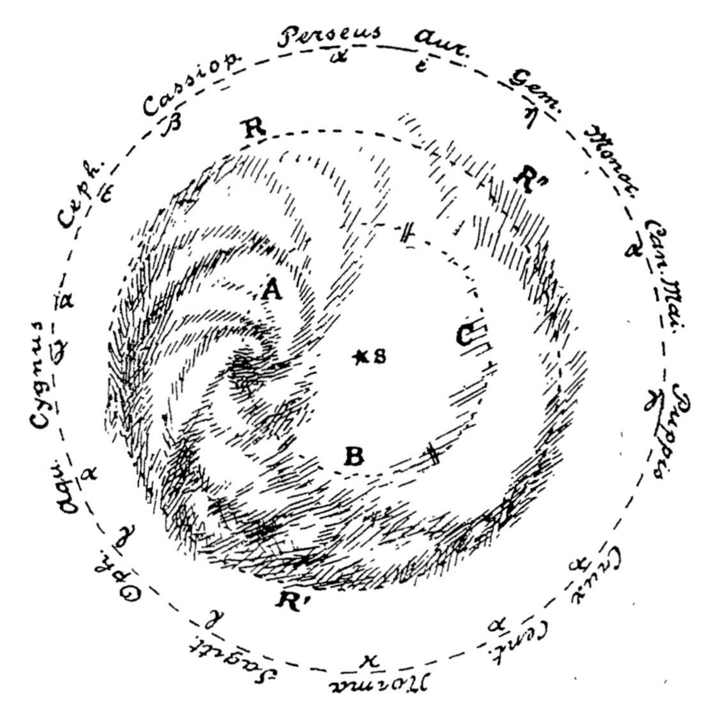

Fig. 3.1 Cornelius Easton's model of our galaxy from 1900. Easton was the first to map the spiral arms of our Milky Way. Source: Easton (1900)

spiral arms, see Fig. 3.1. But there were great uncertainties about the size of our Galaxy and the position of the Sun in it. In 1910, Karl Schwarzschild (1873–1916), an important German astronomer and physicist, concluded from stellar statistics that the Milky Way had a diameter of 10 kpc and a thickness of 2 kpc centered around the Sun (*Astronomische Nachrichten* (*Astronomical Notes*) 1910, 185). Sir Arthur Eddington saw the Sun at a distance of 60 Ly above this disk, but also in the center.

In the astronomical literature, the position of the Sun shifted from the center by several thousand parsecs in the coming years. Thus, in 1911, Kapteyn published a survey of the Milky Way, in which he already moved the Sun 3 kpc out of the center, see Fig. 3.2. Also, the dimensions of the Milky Way increased; 10 kpc diameter around 1900 became almost 20 kpc 20 years later; in short the survey of our star island was anything but complete.

At the beginning of the 1930s, the maps of the Milky Way became increasingly detailed. Thus, the Magellanic Clouds were recognized as extragalactic objects. In addition, several globular star clusters could be identified in the halo (a spherical area around galaxies) of our Milky Way, see Fig. 3.3.

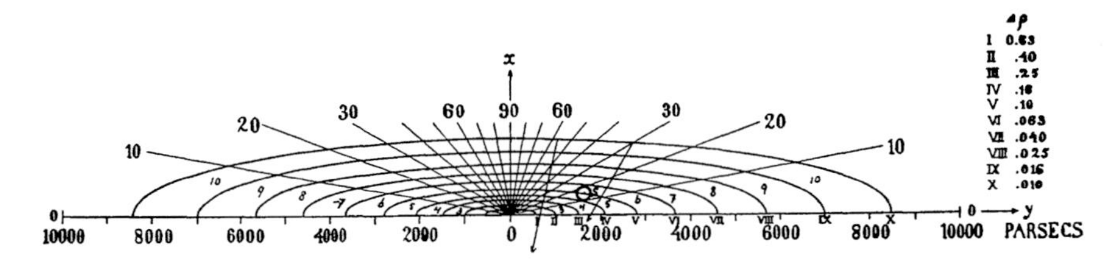

Fig. 3.2 J. C. Kapteyn's model of our galaxy around 1922. The Sun had already moved 3 kpc out of the center. Source: Kapteyn (1922)

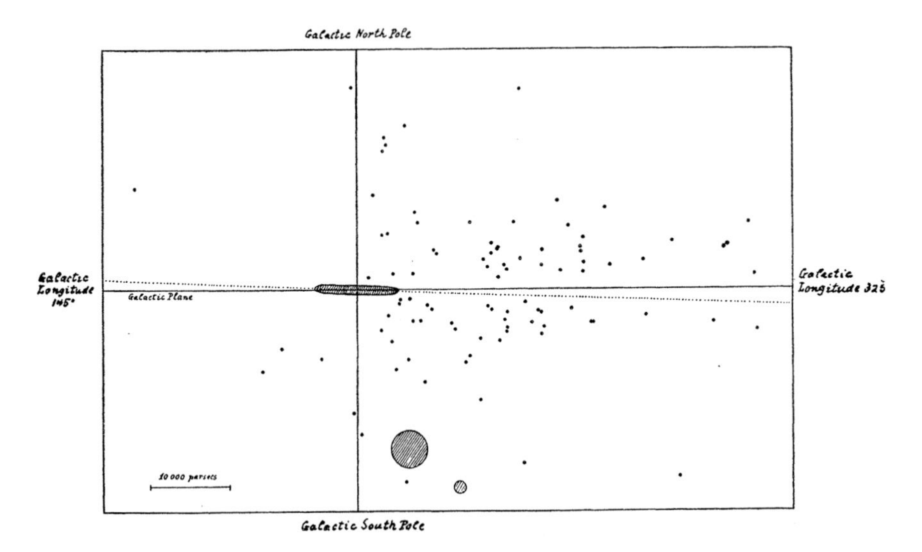

Fig. 3.3 First models of our Milky Way and extragalactic objects by Robert J. Trumpler 1930. Source: Trumpler (1930)

The virial theorem - masses by star distributions

In order to better understand the discovery of a "dark" component of our galaxy by Sir James Hopwood Jeans in 1922, the Milky Way models of a Kapteyn or Easton are sufficient for us at the moment. The distance determination methods used in these models will be discussed in detail later.

The most important and understandable assumption of Jeans was that the stars of the Milky Way are in a state of equilibrium. This assumption, supported by observations, enabled him to estimate the total mass of the Milky Way just by the position of the stars. Jeans used the calculus of the potential theory of gravitation. The basic idea is shown in Fig. 3.4. Since Newton's time, the gravitational force between objects has been described by means of the mass of the bodies involved and their arrangement in space. It can also be formulated in such a way that every body builds up a force field in space, which can be described by the mass and distribution of the bodies. For the sake of simplicity, we want to assume masses as points in space - the bodies have no extension - which may be regarded as a good approximation for the relationship of star radii and typical distances between stars in the Milky Way. The force field resulting from a star exerts a force on every other mass and thus, according to Newton, an acceleration. The force field has the potential to influence the orbits of other bodies. This gravitational potential pervades through space and is only defined by the distribution and masses of objects in the universe. As was the case in Newton's time, the sum of the potentials of the individual masses gives the total potential of the galaxy, see Fig. 3.4. A state of equilibrium is characterized by the fact that - although the mass/stars are moving - global quantities of this gravitational potential do not change on time scales that defy direct observation.

One of these global quantities of a potential is the total mechanical energy in the system. As an example, a planetary system like our solar system is given here, see Fig. 3.5. If one measures the orbital speeds of the planets, he or she will notice that the closer the planet is to the central star, the faster

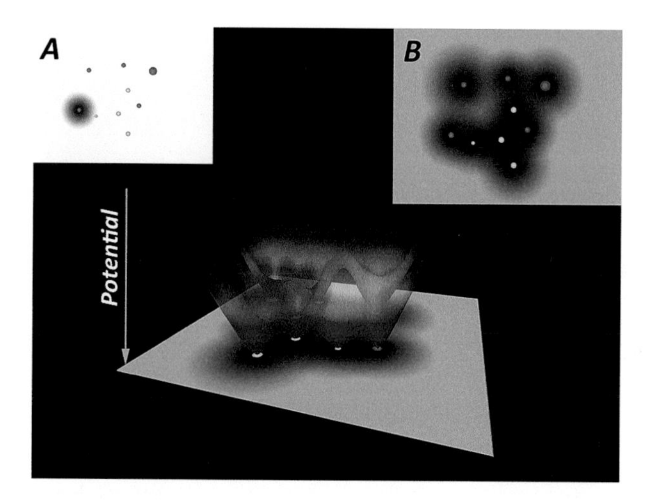

Fig. 3.4 The concept of gravitational *potential* illustrated in two dimensions. Each star (image A) generates a force field around itself. This force field decreases with distance. Its strength is defined by the mass of the star. Image B shows this for some stars, the resulting force field is the sum of the individual force fields. The effect of the force field on a mass in space is shown here by a height map. The redder the area, the closer to large mass concentrations, the more a mass is accelerated

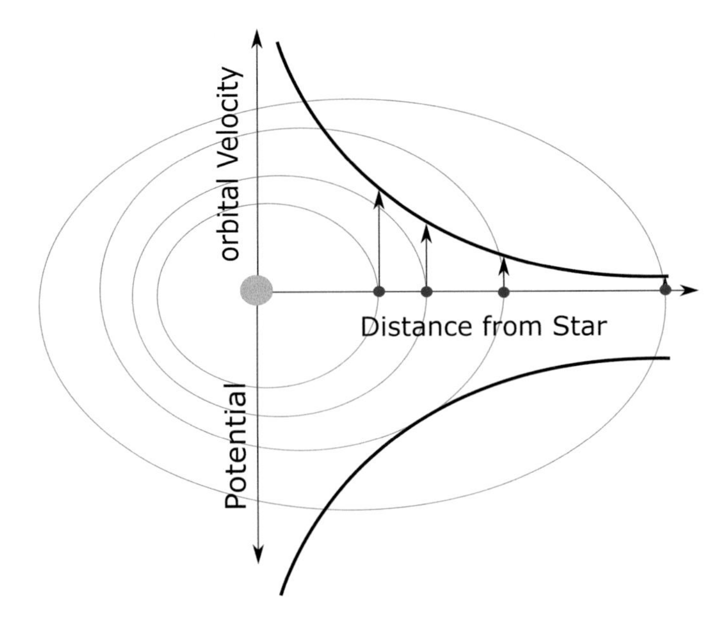

Fig. 3.5 Model of a planetary system around a central star, which dominates the total mass of the system by far. The closer to the central star, the greater the orbital speed of the planets. In addition, the potential defined by the dominating star is shown qualitatively

the planet will orbit. If the potential dominated by the central star is put into this context, a direct connection results.

This connection is known in literature under the term *virial theorem* and was introduced by Rudolf Julius Emanuel Clausius (1822–1888). The energy term used here is to be understood as the movement of objects under the underlying gravitational potential. The most important postulate is: *In every closed system, free from external influences, the total energy is conserved.* Let us consider a stable planetary system. In that system, total mechanical energy is a conservation value. A numerical example of our Earth, which has a mass of 5.972×10^{24} kg and moves around the Sun with an average orbital velocity of 29,780 m/s. It has a kinetic energy of $\sim 2.6 \times 10^{33}$ J, with joules being the unit of energy. A quantity of energy that is not comprehensible to humans and that can only be classified mathematically by comparison with other quantities of energy, which are more accessible to us. This kinetic energy is compared with the potential energy. The potential of a mass as large as that of a star and its planets is approximately a spherically symmetrical potential, as shown in Image A in Fig. 3.4. The orbit of the Earth around the Sun is an ellipse with small eccentricity. Over the course of a year, the distance of the Earth from the Sun increases and decreases slightly. But as the total mechanical energy must be conserved, the orbital speed of the Earth inevitably increases and decreases. Kepler already formulated this very nicely in his laws. The potential energy at the average distance of the Earth from the Sun is about 5.2×10^{33} J, which corresponds to twice the kinetic energy of the Earth in its elliptical orbit around the central star. One can see the elegance of this model, the conversion of kinetic energy into potential energy and vice versa, from the aspect of energy conservation. This is what the *virial theorem* in its core states: *it gives us a relationship between kinetic and potential energy* in a closed, balanced, purely gravitationally interacting system. The question of the total mass of such systems can be reduced to the study of the distribution and velocities of the bodies involved. If these are known, the virial theorem can be used to draw direct conclusions about the total mass.

Let us get back to the groundbreaking work of physicist Sir James Hopwood Jeans. In his model, the stars were lighthouses that, by their arrangement, *illuminated* the gravitational potential and thus revealed to him the total mass of the universe. Of course, the derivation of the potential of all the observed stars is somewhat more complex than in our example of the solar system, but the underlying concept is the same. With this total mass derived from the potential, Jeans now compared the sum of the masses of the observed stars. Motivated by the stellar statistics of those days, Jeans assumed an average of 0.8 solar masses per star counted. And the result astounded him and coming generations of astrophysicists: there was a big difference in the total masses derived in this way. There was simply too little mass in the directly observed stars to reconcile their distribution in the known universe with the theoretical model of gravitation. Jeans trusted the theory of gravity and introduced the term *dark matter* to bring the observations back into line with the theory. He conjectured that they were dark stars and also gave a ratio of visible stars to dark stars. He formulated cautiously in his work "The motions of stars in a Kapteyn universe" (1922 MNRAS 82, page 130):

> Assuming the average star to have a mass of 0.8 times the Sun's mass, our value of M must be interpreted as showing that if the stars are in a steady state, there must be about three dark stars in the universe to every bright star.

For the moment lets capture that Jeans concluded in 1922 that there was 75% dark matter in the universe known at that time. As you will read in the following chapters, this value is not so far away from the value favored today. Unfortunately, however, it was probably too far ahead of its time, as its discovery did not make any big waves. The work has been cited only 59 times since then (as of the end of 2016). The reasons are probably complex, but one must be fair to his colleagues at the time. Much was not yet known or sufficiently researched. For example, Kapteyn's data were published without corrections regarding the absorption of starlight by dust in the galaxy, and there were no mature stellar statistics or stellar models to satisfactorily support assumptions about the mass distribution of stars. And astrophysicists of the time were concerned with quite different questions: What is the nature of the diffuse nebulae whose discovery dates back 150 years? What is the composition of the Milky Way and all the objects observed in the universe? How can the distance-measuring methods be calibrated satisfactorily to each other? Many of these questions could only be answered decades later. However, this in no way detracts from the achievements of the physicist Sir James Hopwood Jeans. One should also quote here the work of the Dutch astronomer Jan Hendrik Oort (1900–1992), who also concluded, through studies of the distribution of stars in the galaxy, that large amounts of dark matter are needed to explain the observed distribution of stars within the framework of the accepted theory of dynamics. In addition, it was Oort who published in 1932, in *Bulletin of the Astronomical Institutes of the Netherlands, Vol. 6,* that there is evidence that this invisible matter must be more concentrated toward the galactic disk than the distribution of the observed stars.

3.2 Early Signs in Galaxy Clusters

Shortly after the early signs of dark matter in our Milky Way through Jeans and Oort, Fritz Zwicky (1898–1974) succeeded in 1933 in uncovering another clear indication of a discrepancy between the mass motivated by the dynamics of galaxies in a galaxy cluster and the mass derived from luminous matter. Zwicky was a native Bulgarian who spent his school and university years in Switzerland before moving to the United States in 1925, where he conducted research in atomic physics, astronomy and space travel. Zwicky was the first Western scientist to launch a body into space 12 days after the first artificial satellite, the Russian Sputnik I. According to various sources, he was not only a controversial contemporary, but also a universal genius of his time. In order to understand Zwicky's line of argument, we must look at the early discoveries in the field of extragalactic astronomy in the 1920s and 1930s. In particular, we need to look at the large-scale distance determination methods that revolutionized the picture of the then known universe.

Distance Measurements Beyond the Boundaries of Our Galaxy

Since the middle of the eighteenth century, the lists of objects observed by astronomers had already included diffuse objects that differed greatly from the point-like images of stars. These objects were comets or clusters of stars, so-called star clusters, planetary nebulae, repelled matter from stars in their final stages of development, or simply luminous gas clouds in our Milky Way. But many of the nebulous objects mapped at that time were not yet recognized galaxies of various shapes and sizes outside our galactic home, hence the term extragalactic. One of the most famous of these object catalogs was and is the Messier catalog, named after its author, the French astronomer Charles Messier (1730–1817). It contains about 100 identified astronomical objects, most of them are galaxies or star clusters and gas nebulae in our Milky Way. For example, our neighboring galaxy in the universe - *Andromeda* – has the number 31 and is nowadays simply listed in the catalog as M31 after Messier's catalog. Another famous example is the Sombrero galaxy with the designation M104, see Fig. 3.6.

The true nature of these nebulous objects was recognized with the advent of the first large telescopes around 1910. It were the powerful american

Fig. 3.6 Messier catalog object 104, the Sombrero galaxy, observed with the Hubble Space Telescope (M104). Picture credits: NASA and the Hubble Heritage Team (STScI/AURA)

reflector-telescopes of those days that brought the breakthrough in the exploration of these objects. A 1.1-m reflecting telescope was installed at the Lowell Observatory in Arizona in 1910 and a 2.5-m reflecting telescope was installed on Mount Wilson in California in 1917. More than a decade later, the reflector-telescopes of the Palomar Observatory in southwestern California followed. It had been known for some time that many of these objects were collections of stars, some similar to our own Milky Way, but it was not yet certain whether these objects themselves were part of our Milky Way or represented galaxies of their own. What was missing was a reliable method for measuring distances. This changed with Edwin Powell Hubble (1889–1953), an American astronomer who found a relationship in the collected data that would take astronomy into a new era. At the beginning of his career, he successfully tried to measure the distance to M31 and M33, two nebulae already known at the time of Messier, using a method called Cepheid measurements, which was calibrated within the Milky Way.

The Cepheid Method

A young English astronomer named John Goodricke (1764–1786) had accurately measured the brightness fluctuation of a star called Delta Cephei and discovered that the brightness of this star is related to a period of about $5\frac{1}{4}$ days, see Fig. 3.7. However, it was not until 125 years later that the American astronomer Henrietta Swan Leavitt (1868–1921) systematically investigated the brightness fluctuations of such stars in the Magellanic Clouds, which are dwarf galaxies from today's perspective, trapped in the gravitational potential of our Milky Way. It was their achievement to realize that there is a correlation between the period length and the magnitude

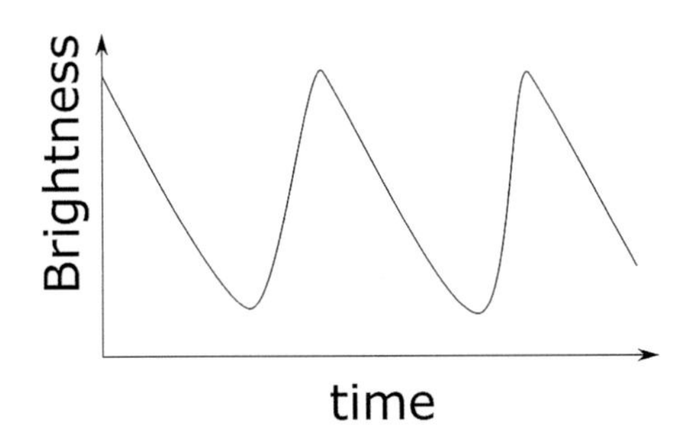

Fig. 3.7 Typical brightness curve of the Delta Cephei variable. From maximum to minimum typically a few days pass

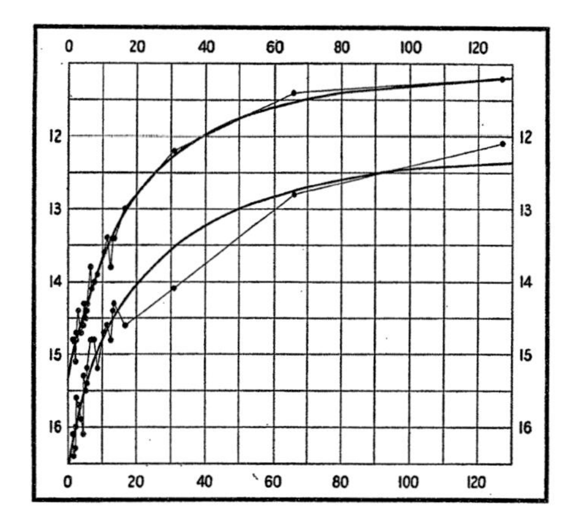

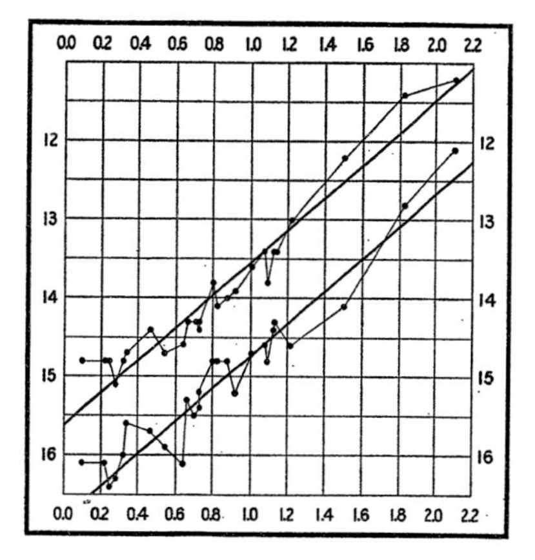

Fig. 3.8 The period–luminosity relationship in Cepheids published by Henrietta Swan Leavitt. The left side shows on the abscissa the period duration in days and on the ordinate the apparent brightness, each for the maximum and the minimum luminosity. The right side shows the same as the left side, but this time with a logarithmic time axis. Source: Leavitt, Pickering (1912)

of the brightness fluctuation. This finding forms the basis for the distance method known today as the Cepheid method.

Figure 3.8 shows Leavitt's original graph of the period–luminosity relationship of the Cepheids. These are periods and magnitudes of 25 variable stars in the small Magellanic Cloud, which she published in 1912. The left side shows on the abscissa the period in days and on the ordinate the apparent magnitude, each for the maximum and the minimum of the measured luminosity of the variable star. The right side shows the same as the left side, but this time with a logarithmic time axis. She found that the average difference between the minimum and maximum brightness of the 25 stars is a constant 1.2 magnitudes and that if the period of change increased by a factor of 0.48, the brightness of the star increased by 1 magnitude. But an important piece of the puzzle was still missing. One needed the absolute magnitudes of the stars to be able to use them as distance meters. The idea was the following: If one could generally determine how bright these variable stars are in absolute terms, then one could simply measure the distance to these kinds of stars by measuring a period. Moreover, Cepheids are very bright stars, so they would be ideal candidates for measuring the distance to our neighboring galaxies.

It was Ejnar Hertzsprung who in 1913 attempted a first calibration of the Cepheids observed by Leavitt. Hertzsprung used a star catalog called *Preliminary General Catalogue of 6188 Stars* by the American astronomer

Lewis Boss (1846–1912), in which 13 stars were found as Delta Cephei variable stars with information on the periods, the maxima and minima of the magnitudes and their proper motion relative to the Earth, the so-called secular parallax. The secular parallax is caused by the fact that our Sun, with all its planets, moves with about 20 km/s in the direction of the constellation Hercules. In observations of the positions of the stars over the years, this causes an apparent movement of the stars in the opposite direction to us. However, this parallax can only be applied statistically, over many stars, and gives us a quite good distance determination. And this was exactly the approach of Hertzsprung to determine the absolute magnitudes of these Cepheids found by Boss. He used this apparent motion of the stars to deduce the distance of these 13 Cepheids. This represents the core of the idea of a distance ladder in astronomy.

You have solid, well-studied distances for the nearby stars, the first rung on our ladder. You find a correlation, for example, the period and the difference in brightness of some similar variable stars. You look for comparable stars, of which you think you know the distance well based on established methods, and you calibrate their absolute magnitudes. You are already on the second rung of the ladder. We will see with which methods one can climb the ladder further. It is important to emphasize that if there are errors in distance determination on the lower rungs, the error will propagate and our distance determination ladder will become shaky and collapse like a house of cards.

Back to Hertzsprung again: He now came to an absolute brightness for Cepheids with a period of 6.6 days of -7.3 magnitudes So he was able to calculate a distance of 30,000 Ly for the Cepheids of the same period found by Leavitt in the small Magellanic Cloud and so it was clear that this cloud is outside the disk of our Milky Way. It was the first step in extragalactic astronomy. Unfortunately, as it turned out later, the distance determination of Hertzsprung was incorrect. On the one hand, it was subject to a conversion error, and on the other hand, the data of its reference stars were not good enough. But the idea, the approach was the outstanding thing. With current data and better calibrations of the distance scale, the distance to the small Magellanic Cloud is now estimated to be about 209,000 Ly.

What causes these fluctuations in brightness? One understands the brightness fluctuation of such stars as a cycle in which a certain layer in the stellar envelope changes its temperature and pressure by interacting with the light from the star's interior. This also changes the light transmission of this layer, the brightness. This, in turn, changes the pressure and temperature of this layer of stellar envelope as it emits energy in the form of radiation,

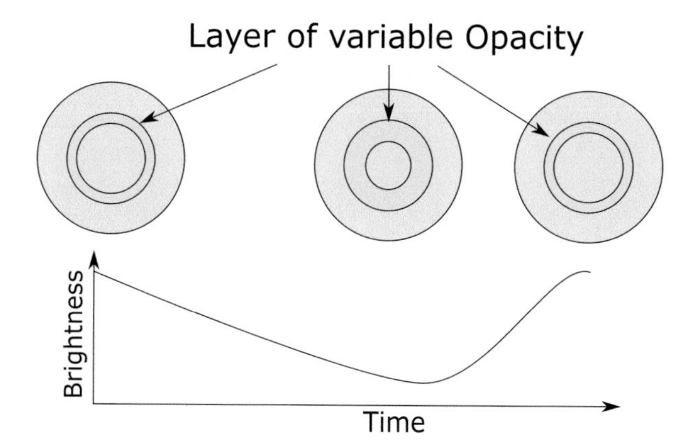

Fig. 3.9 Schematic diagram of the κ-mechanism. Due to pressure and temperature stratifications in the star envelope, light from the interior of the star may have a different light transmission in a certain region. As matter and light interact, the temperature and density in this layer change again and so does the light transmission. A pulsating cycle is created, which is reflected in the change in the brightness of the star

which again changes the light transmission - a cycle that well describes the brightness fluctuations of variable stars, see Fig. 3.9. This mechanism is called the κ-mechanism after the Greek letter kappa (κ) used in literature for the light transmission of a material.

In many variable stars, this process causes the star to pulsate, which can also be studied well using spectroscopic observations.

With current generations of telescopes, like the Hubble Space Telescope, Cepheids can be observed up to a distance of about 20 Mpc, and thus Cepheids serve as *standard candles* to determine the distance up to the largest cluster of galaxies closest to us, the Virgo cluster of galaxies. Besides the Cepheid stars, there are several other well studied periodically changing star types.

But we come back to Hertzsprung, who determined the distance to the small Magellanic Cloud with the help of the Cepheids and thus showed that this nebula must be located far outside the disk of the Milky Way. Not only Hertzsprung tried this method, but the young astronomer Edwin Powell Hubble also determined the distance to spiral nebulae M31 and M33 with observations from the most powerful telescopes at Mount Wilson Observatory. He published the distances found in 1925, giving 285,000 pc for both spiral nebulae. Today, the distance to M31 is given with about 766,000 pc and to M33 with about 850,000 pc. Again, it was absolutely clear that these were extragalactic systems. In Fig. 3.10, one of the Cepheid

Fig. 3.10 The Cepheid variable star V1 in the Andromeda galaxy M31, observed with the Hubble Space Telescope. Picture credits: NASA, ESA and the Hubble Heritage Team (STScI/AURA)

variable stars that Hubble used (V1 in M31) is shown. The observations were made with the Hubble Space Telescope, the telescope that bears the name of that great astronomer.

However, Hubble became famous for a completely different distance determination method, namely the relationship between the measured radial velocity and the distance of extragalactic nebulae, a method that opens the door to distance determination into the depths of the cosmos. In astronomy, radial velocity is the velocity component of a movement in the direction of observation - either negative in the direction of the observer or positive in the direction away from the observer.

The Redshift of the Spectra

The shift of lines in the spectra of stars was first systematically studied in the mid-nineteenth century by Sir William Huggins (1824–1910), a British

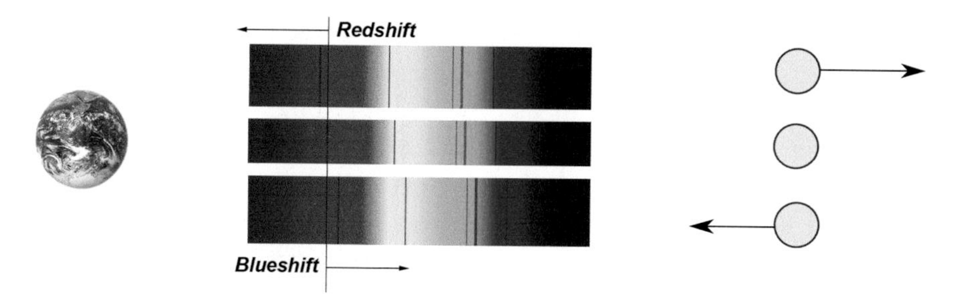

Fig. 3.11 Three stars are observed: one star moves toward the Earth, another star is at rest in relation to the Earth and a third star moves away from it. The spectrum is observed in relation to the resting star with one redshifted (star moving away from us) and one blueshifted (star moving toward us)

astronomer and physicist. In 1868, he wrote in a publication,[1] among other things, about the shift in the wavelength of a strong hydrogen absorption line in the spectrum of light from Sirius. He also calculated a velocity of Sirius relative to Earth and came up with a value of about 65 km/s, a value that can be considered a great observation success at the time of Huggins, because during his lifetime, it was not yet known what causes the shift of the lines in a spectrum. For Huggins, the only conceivable cause was Sirius' flight from Earth. Today, this effect is known as the relativistic Doppler effect, named after the Austrian physicist and mathematician Christian Andreas Doppler (1803–1853), who became famous for his description of the classical Doppler effect in acoustics. This effect is shown schematically in Fig. 3.11. Three stars are observed: one star moves toward the Earth, another star is at rest in relation to the Earth and a third star moves away from it. The positions of the absorption lines in the spectrum of the star moving toward Earth shift toward shorter wavelengths (blueshift) compared to those of the resting star. The spectrum of the star moving away from the Earth, on the other hand, is observed redshifted - toward longer wavelengths. If we assume that a light source emits a line at exactly 500 nm and it moves away from us at 1 km/s, we measure a wavelength of 500.002 nm.

[1]Huggins W (1868) Further Observations on the Spectra of Some of the Stars and Nebulae, with an Attempt to Determine Therefrom Whether These Bodies are Moving towards or from the Earth, Also Observations on the Spectra of the Sun and of Comet II. Philosophical Transactions of the Royal Society of London 158.

However, the first measurements of spectra of the nebulous objects had to wait almost 50 years. It was not until around 1910 that telescopes and spectrographs were able to fan out the light of some nebulae such as the Andromeda Nebula M31 into its wavelengths/frequencies. This was first achieved by the American astronomer Vesto Melvin Slipher (1875–1969) at the Lowell Observatory in Flagstaff, Arizona. In 1912, Slipher was able to determine the mean velocity from M31 with respect to us in several measurements with the value − 300 km/s. This means that the Andromeda galaxy M31 is moving toward our Milky Way at 300 km/s. He was very surprised by this discovery himself and assumed that even under the assumption that M31 is part of our Milky Way, it would interact with an invisible star that could cause such a high, previously unobserved speed. As it turned out, Slipher was wrong about this *dark star*. Hubble's observations 10 years later made it clear that it is an independent galaxy far outside the Milky Way, approaching it at 300 km/s. But the measurements by Slipher paved the way to one of the most important relations in extragalactic astronomy, which was published 17 years later in 1929 by Edwin Hubble: the *velocity–distance diagram of extragalactic nebulae*.

Hubble had spectra of 46 extragalactic nebulae from which he could measure radial velocity. In addition, he collected distance data to these nebulae, partly from "robust" methods such as the Cepheid method, and partly by estimating the absolute magnitudes of the brightest stars in these nebulae. Thus, he could correlate the two quantities - a radial velocity and a distance - per nebula and lo and behold, there seemed to be a correlation. The original correlation is shown in Fig. 3.12. Today, one could say that Hubble was very "bold" in his original work. There were only a few measuring points, strong assumptions and many approximations involved, but Hubble's courage was rewarded because further investigations supported his initial assumptions.

But what had Hubble found here? The further away from us a galaxy is located, the faster it moves away from us. This means conversely, if you can measure the shift of the spectral lines of extragalactic nebulae against resting reference lines accurately, you have a measure of the distance of the galaxies. The slope of the best fit line is called Hubble constant, it is the measure of how much the velocity increases with distance.

Let us recall: In the early 1930s, there were strong evidence that there was a lot of invisible matter in the Milky Way. And now a new estimation of masses came into play, a long-range estimation, and this was done by a universal genius of the last century, Fritz Zwicky. To understand this estimation, we have introduced all the arguments.

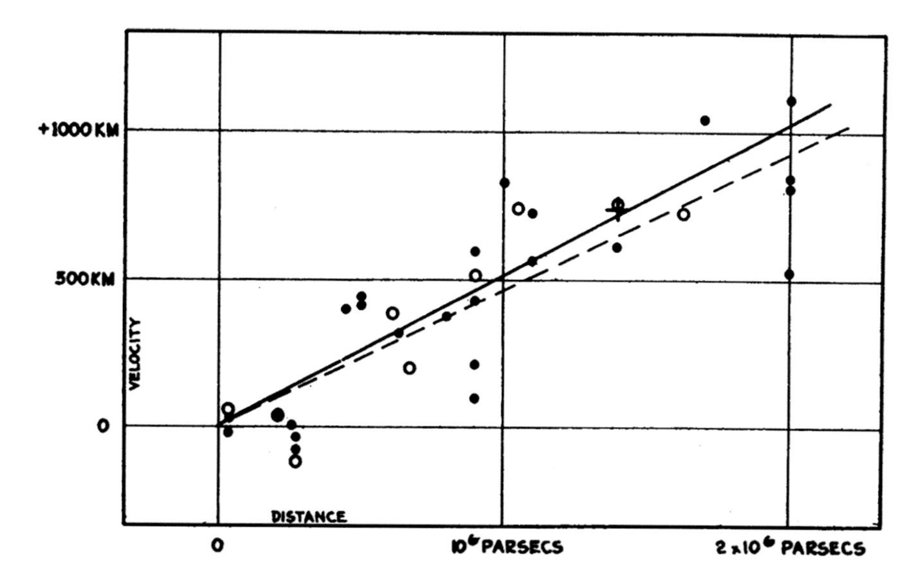

Fig. 3.12 Velocity–distance diagram of extragalactic nebulae from the original work "A Relation between Distance and Radial Velocity among Extra-Galactic Nebulae" by Edwin Hubble (1929), published in the Proceedings of the National Academy of Sciences of the United States of America, Volume 15, Issue 3. A few nebulae move in the direction towards the Milky Way (negative radial velocities), the large majority have positive velocities. The slope of the compensation line is called Hubble constant, it is the measure of how much the velocity increases with distance

Zwicky and the Coma Cluster of Galaxies

In 1933, Fritz Zwicky published a work entitled "The Redshift of Extragalactic Nebulae". It was kind of an overview summarizing the state of knowledge at these days. The main part of the work was dedicated to the different ways of determining the distance of extragalactic objects. He mentioned, for example, the Cepheid method and the method of statistics of stars of highest brightness in a nebula. These two approaches had been used in the early 1930s to determine the distances to about 60 extragalactic nebulae. By investigating the distribution of these nebulae, Zwicky showed that the apparent spatial clustering of extragalactic objects could not only be a projection randomness in the sky, but that they must probably be accumulations of galaxies, so-called galaxy clusters. He recognized that the number of nebulae per unit volume in these clusters is at least 100 times greater than the average number of nebulae averaged over the entire universe known at that time. He also made considerations about the existence of intergalactic matter in the form of gases and microbodies (dust) that could be located between the observed nebulae and thus would have to weaken the light emitted by the galaxies. As he did not detect any systematic decrease in the brightness of galaxies with distance caused by intergalactic matter using the 100-inch giant telescope at the Mount Wilson Observatory, he concluded that there are no significant amounts of this intergalactic matter. Due to the enormous light-collecting power of the 100-inch telescope, Zwicky was able to obtain very useful spectra of these extragalactic nebulae and he systematically studied them. He succeeded in showing that the nebulae have absorption spectra similar to those of the Sun. Furthermore, he found no change in the spectral type with the distance of the nebulae. He also found emission lines that came from the central region of the nebulae, probably the first observations of active galaxy nuclei. Zwicky found that 74% of all the galaxies observed at that time had spiral morphology, 23% were spherical and about 3% had irregular appearance.

He also gave a brief overview of Edwin Hubble's work on the redshift of extragalactic nebulae and showed how challenging the observation of distant nebulae was for spectroscopy at that time from a technical point of view. Zwicky also derived, based on the cosmological models of his time, an average density of the universe of about 10^{-28} g/cm^3. With regard to the cosmological models of his time, Zwicky summarized that all of them were developed on an extremely hypothetical basis and revealed practically no new physical relationships.

But the penultimate chapter of his work contains a key observation for the phenomenon of dark matter. It is entitled "Remarks on the scattering of velocities in the Coma nebula cluster". What did Zwicky notice?

At the time of Zwicky, there were about 800 known galaxies in the Coma galaxy cluster. The galaxy cluster is observed in the constellation Hair of the Berenice (Coma Berenices), a constellation between Leo and Boötes, and is therefore called Coma, see Fig. 3.13. Zwicky had measurements of the redshift of some cluster members and discovered that the radial velocities derived from them scatter around the mean radial velocity (about 7,000 km/s) of the galaxy cluster at about 1,500–2,000 km/s. This aroused Zwicky's interest, he was able to derive a mass for the entire galaxy cluster, which he succeeded by using the virial theorem. He assumed that the Coma system was in a state of equilibrium and that the mass was evenly distributed across the cluster. In addition, he determined the radius of the Coma Cluster at 1 MLy and estimated the mass of a galaxy with 1 billion solar masses. Thus, Zwicky was able to estimate the total potential energy of the Coma Cluster and equate it with the mean kinetic velocity according to the virial theorem. In a purely gravitationally interacting system, which is in a

Fig. 3.13 The Coma cluster of galaxies observed with the Hubble Space Telescope. The galaxy cluster is observed in the constellation Hair of the Berenice (Coma Berenices), a constellation between Leo and Boötes, and is located at a distance of about 300 MLy. Source: NASA, ESA, and the Hubble Heritage Team (STScI/AURA)

state of equilibrium, the potential and kinetic energy are linked. Zwicky calculated approximately from this that the difference in velocity of the galaxies around the mean velocity in the Coma Cluster could not be greater than 80 km/s. He concluded that the mean density in this galaxy cluster needed to be 400 times greater to obtain the observed mean velocity dispersion of around 1,000 km/s. He writes: "If this were to prove true, the surprising result would be that dark matter exists in much greater density than luminous matter." He then estimated the mass again for the extreme case that the entire mechanical energy of the galaxy cluster is in the motion of the galaxies, which would weaken the previous estimate by a factor of 2. In addition, he was looking for another way out, he assumed in a thought experiment that no dark matter existed. For the Coma Cluster, this means that it would have to fly apart. Now he searched for galaxies, away from galaxy clusters, so-called field galaxies, and examined their velocity scattering. These would have to have an average dispersion of more than 1,000 km/s. However, as a typical scattering of 200 km/s is observed for field galaxies, this means that galaxy clusters are bound systems - not randomly clustered arrangements of galaxies. In this model building, there must therefore be much more dark matter in galaxy clusters than visible matter.

Zwicky's considerations regarding the missing mass were not considered further for many years. But this changed in the mid-1980s, as can be seen very clearly from the number of citations of this work per year in Fig. 3.14. Certainly, the assumptions, which were only weakly supported by observations, played a major role. Certainly, other questions were more urgent to investigate, questions about the nature of these nebulae, questions about the cause of the observed redshift. But Zwicky's approach was important because he showed in a very elegant way that there is a difference between the masses determined by the dynamics and the luminosity of objects, not only in our Milky Way but also in the clusters of galaxies in the cosmos.

But let us return to the estimates of Sir James Hopwood Jeans in our Milky Way, who found a difference of a factor of 3 with the same calculation, and note that only 3 years later, a factor of more than 200 of dark matter was already needed in galaxy clusters.

Perhaps Sir Arthur Stanley Eddington was very right in his statement:

> … It is good advice not to put too much weight into a theory until it has been confirmed by observation. But I hope not to shock experimental physicists too much when I add that it is also good advice not to put too much weight in the result of an observation until it has been predicted by a theory.

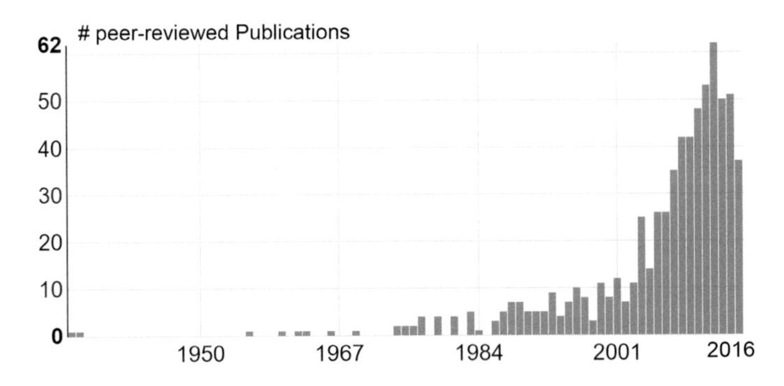

Fig. 3.14 Number of citations of the publication "The Redshift of Extragalactic Nebulae" by Fritz Zwicky from the year (1933), published in the *Helvetica Physica Acta*, in peer-reviewed publications. Source: The SAO/NASA Astrophysics Data System (ADS)

Summary

We learned in this chapter essentially about two early signs of dark matter: first, in our Milky Way through the work by Sir James Hopwood Jeans, and second, in the collections of galaxies, so-called clusters of galaxies, by the work of Fritz Zwicky. It is important to note that in both publications, there is a large difference between the masses derived from the dynamics and the luminosity of the objects. In all these investigations, the determination of the distance to the objects plays a central role. This chapter introduced the concept of distance ladders in astronomy, the calibration of different distance determination methods to each other. The result changed dramatically our view of the universe at the beginning of the twentieth century and showed us that the Milky Way is only one galaxy among many.

But the dark matter phenomenon did not disappear in the following decades. On the contrary, it increasingly dominated discoveries on large scales. The next chapter will show how dark matter became more and more widespread in the world of astrophysics and how recent observations and interpretations have refined our knowledge of its properties.

References

Easton C (1900) A new theory of the milky way. ApJ 12:136–158

Hubble E (1929) Proc Natl Acad Sci USA 15(3)

Kapteyn JC (1922) First attempt at a theory of the arrangement and motion of the sideral system. ApJ 55:302–328

Leavitt HS, Pickering EC (1912) Periods of 25 variable stars in the small magellanic cloud. HC 173:1–3

Trumpler RJ (1930) Preliminary results on the distances, dimensions and space distribution of open star clusters. LOB 420:154–187

Zwicky F (1933) Die Rotverschiebung von extragalaktischen Nebeln. Helv Phys Acta 6:110–127

4

A Problem Manifests Itself

Already at the beginning of the twentieth century, a problem in the mass budget of large structures like our galaxy or galaxy clusters became obvious. The mass derived from the brightness of the objects was different from the mass derived from the theory of dynamics - not just a little, but exorbitantly. But the observations made at that time were not good enough to be fully believed. Thus, several years passed before a large part of the astrophysical community was fully engaged with the implications. It was the time when the problem of *missing mass* became established.

The research work of those years was closely linked to developments in the field of observation technologies. New areas of the electromagnetic spectrum were constantly being made accessible for observational astronomy, thus opening up ever wider windows into the cosmos. And what there was to discover shook the world view of astrophysicists again and again. The new physical models, such as the general theory of relativity or quantum mechanics, had already proven their validity in many fields several times. In addition, they were able to satisfactorily describe many astrophysical observations of those days. It was therefore not surprising that the phenomenon of dark matter was regarded as a temporary problem at the time, a phenomenon that could be revealed as trivial with ever improving observation technologies.

But this belief was not fulfilled, quite the contrary. With new observations, the problem of dark matter became a bigger one. In this chapter, you will be introduced to the developments of that time on the basis of some key observations and in addition the physical models involved will be explained. Models and observations that led to a universe hidden in darkness.

W. Kapferer, *The Mystery of Dark Matter*, Astronomers' Universe,
https://doi.org/10.1007/978-3-662-62202-5_4

4.1 Rotation Curves of Spiral Galaxies

There has been speculation about the possible rotation of nebulae, especially spiral nebulae, since their discovery. However, it was not until 1914, in a publication by Vesto Melvin Slipher, that this speculation was substantiated by measurements. Slipher examined the spectra of the galaxy M104, see Fig. 3.6, and recognized a shift of the spectral lines in such a way that only a rotation of the disk could be a possible explanation. But it was not until 1939 that Horace Welcome Babcock (1912–2003), an American astronomer, studied the rotation of the Andromeda galaxy M31 in more detail and compared these data with the observed mass distribution of this galaxy. He published his investigations in the article "The rotation of the Andromeda Nebula," published in *Lick Observatory bulletins* (1939). In Fig. 4.1, the main observation of the work is shown in original. Babcock investigated the radial velocity extracted from spectra at different positions of the Andromeda galaxy M31. From this, he was able to create a so-called *rotation curve*, a velocity curve of the objects along the observed disk of the galaxy.

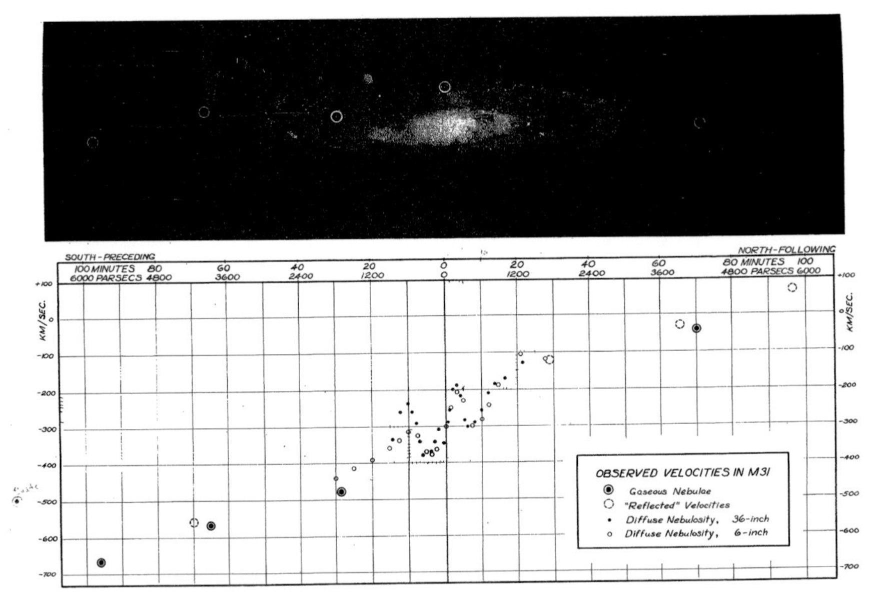

Fig. 4.1 Original image of the observed radial velocities of the Andromeda spiral galaxy M31. For different positions in the galaxy, spectra were recorded, their line shifts were analyzed and the radial velocities derived from them were plotted. Source: Babcock (1939)

This rotation curve was far from being consistent with the observed mass distribution in Andromeda.

Babcock's interpretation of this rotation curve was groundbreaking. He modeled a mass distribution of the galaxy based on its measured dynamics that would fit the rotation curve shown in Fig. 4.2. The central consideration was to divide the galaxy into four regions: a central spherical region (A) and three other very flattened ellipsoids (B, C and D). This allowed him to assign a rotational speed to each region from the measured rotation curve and to derive how large the mass in the different regions would have to be in order not to violate the laws of dynamics. As a result, Babcock obtained a total mass for M31 of about 100 billion solar masses. In the next step, he derived the mass of the galaxy using the apparent brightness of the stars. From the known distance, he was able to determine the mean absolute brightness of the entire galaxy, which corresponded to a luminous mass of about 2 billion suns. He also applied this method to the four regions of the galaxy. Again, he calculated the ratio of luminous mass to dynamic mass in these four regions and was able to show that the further away from the core of the galaxy, the lower the ratio of luminous mass to dynamic mass. In the outermost region, there was about 60 times more dark mass than luminous mass. His interpretation of this result was very careful. He summarized:

- Either the absorption of the light emitted by the stars by gas and dust within the galaxy plays a very important role and thus the overall brightness of the galaxy is lower.
- Or it needs new considerations in the field of the dynamics of the motion of stars.

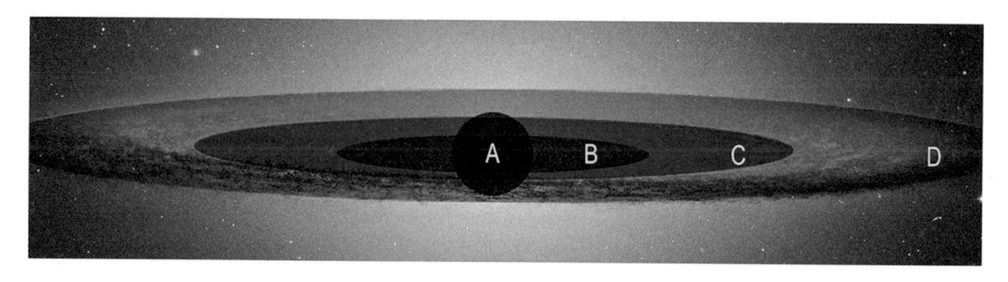

Fig. 4.2 Horace W. Babcock's mass distribution model for M31, schematically shown. He divided the galaxy into a central sphere and three further flat ellipsoids, for each of which he determined the mean rotational speed and was thus able to derive their dynamic mass. The result was about 100 billion solar masses for the entire Andromeda galaxy

Babcock was not averse to modifying the fundamental theory of dynamics.

I guess Babcock's discovery came too soon. The astrophysicists of those days were mainly concerned with describing the different shapes of galaxies in models that were as simple as possible and with few parameters. Such a classification of the different forms of observed galaxies was made by Edwin Hubble in 1926 and entered the literature as the *Hubble sequence*. However, this classification of galaxies from elliptical to spiral and barred spiral galaxies had to be constantly expanded and adapted, because with the advent of ever better telescopes and the resulting observations, these objects revealed more and more details. In addition, attempts were made to derive a kind of evolution of the galaxies from their shapes. In short, the general tenor was that the limited observations had simply not yet revealed all the matter in the outer regions of the galaxies. This assumption was also fueled by the astonishing results of the new observational possibilities emerging on the horizon in the regions of the invisible electromagnetic spectrum. First observations were able to partially wrest darkness from dark matter. For example, the British radio astronomers Hanbury Brown (1916–2002) and Cyril Hazard were the first to detect radio radiation outside our Milky Way in the Andromeda galaxy M31 in 1950. Figure 4.3 shows contour lines from their original work "Radio emission from the Andromeda nebula," published in *Monthly Notices of the Royal Astronomical*

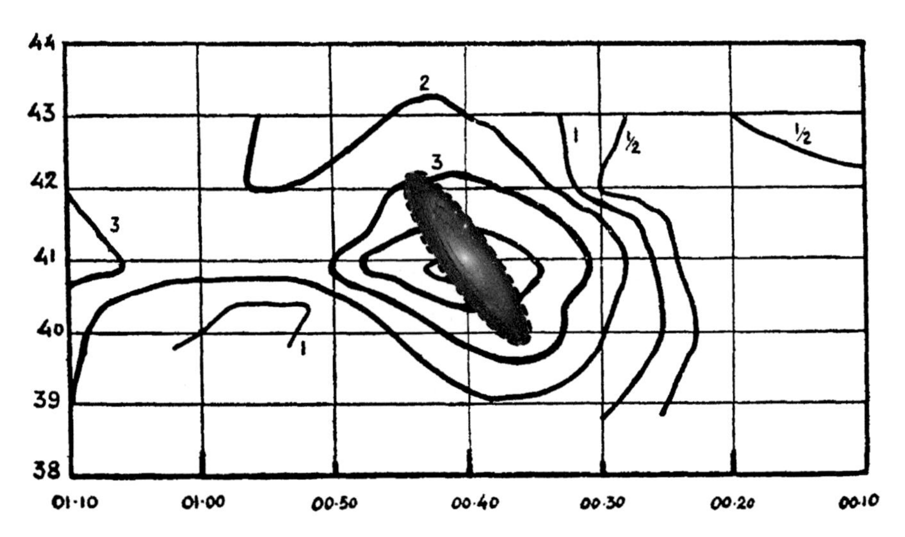

Fig. 4.3 Radio radiation from the Andromeda galaxy M31, located outside our Milky Way, observed by Hanbury Brown (1916–2002) and Cyril Hazard. Contour lines from their original work "Radio emission from the Andromeda nebula," published in *Monthly Notices of the Royal Astronomical Society* (1951), are shown. For size comparison, an optical image of the Andromeda galaxy M31 is shown

Society. For size comparison, an optical image of the Andromeda galaxy M31 is shown. It can be seen that there is matter far beyond the optical boundaries of M31. The origin of this long-wave (1.89 m) electromagnetic radiation was unknown at that time. Nevertheless, it was clear that there is far more matter in the outer regions of a galaxy than optical telescopes had previously revealed.

Although questions about the dynamics of these systems and the masses derived from them have been part of individual research projects ever since this early work by Babcock, it was not until the late 1970s that a critical mass of data was collected in order to grow the conviction within the astrophysical research community that there is a large discrepancy between the directly observed masses and the masses derived from the motion of objects in galaxies. The studies of the American astronomer Vera Cooper Rubin (1928–2016) were decisive. Rubin and her colleagues were able to determine rotation curves of several galaxies through numerous observations from the end of the 1960s onward, and repeatedly found that the rotational speeds in the outer regions of the galaxies did not decrease as expected. The measurements of galaxy rotations in the radio range also showed this behavior (Van de Hulst et al. 1957). It was therefore clear at this point in time that there was a large gap in the knowledge of the mass components of galaxies. In Fig. 4.4, this is shown by the measured rotation curve of the Andromeda galaxy M31. In white, observed data and a best fit from a paper by Vera Rubin

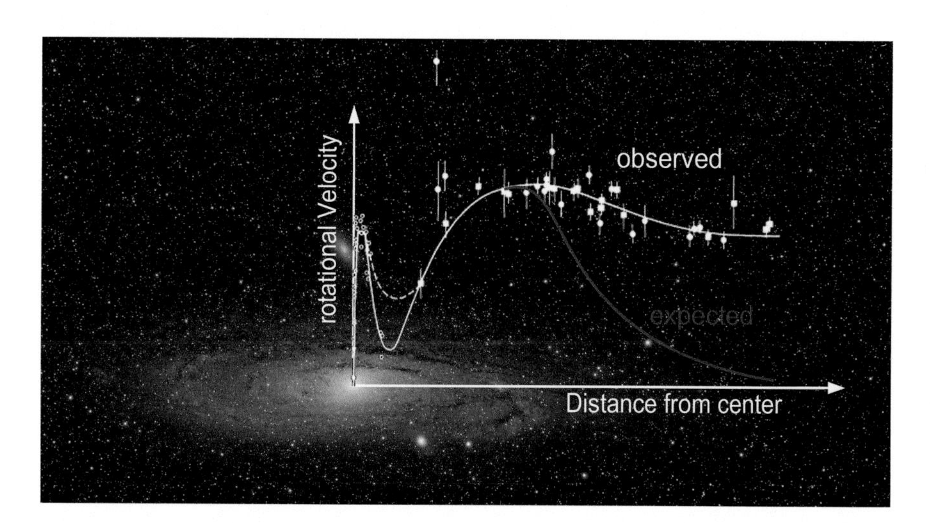

Fig. 4.4 Rotation curve of M31. In white, the observed data and a best fit from a work by Vera Rubin and Kent Ford (1970) are shown. In red, the expected curve of the rotation based on the observed mass distribution according to classical dynamics is shown

and Kent Ford, published in *Astrophysical Journal*, (1970), are shown. The red line corresponds to the expected course of the rotation curve in the outer region, that is, if the directly observable mass distribution - corresponding to the brightness of the different mass components, such as stars, gases and dust particles - and Newtonian dynamics would describe the galaxy correctly.

Observation and model are strongly contradictory at this point. In order to better understand this discrepancy, let us recall the starting point of the question of how masses are determined by the dynamics of purely gravitationally interacting systems. The underlying measurement data for an important test of Newtonian dynamics come from the orbits of the planets in our solar system, the Kepler orbits, as shown in Fig. 4.5. Let us note again that in our solar system, almost all mass is concentrated in the center. This means that the strongest forces and thus the greatest accelerations occur in the vicinity of this center of mass. Therefore, the high orbital speeds of the inner planets in comparison to the outer planets can be well described. With galaxies, the observable mass decreases toward the outside, so one would expect lower accelerations and orbital speeds in the outer regions according to the theory of dynamics. However, this is not the case. As the observed rotation curves of galaxies in the outer regions do not become flat, as is the case with the planets in the solar system, two conclusions are possible:

- The *observations* in the outer region of galaxies are incomplete or
- The celestial mechanics based on Newton's theory of gravity fails in these regions of the galaxies.

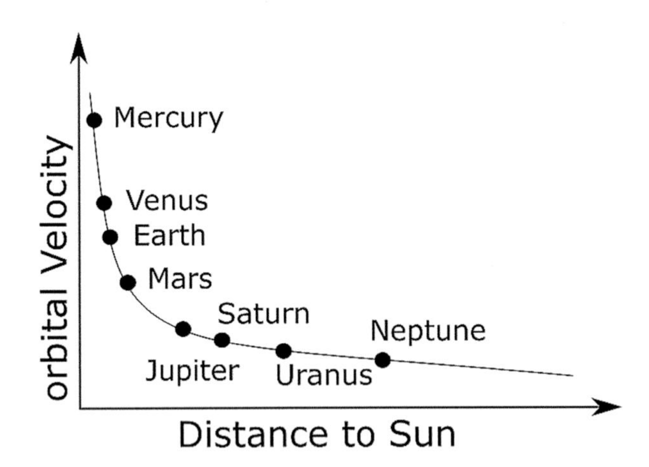

Fig. 4.5 "Rotation curve" of the planets of our solar system. The orbital speed is getting slower and slower from the inner to the outer planets

At this point, let us recall the history of the exploration of binary stars in Chapter 2. In 1844, Friedrich Bessel recognized that Sirius is performing a peculiar motion on the firmament. At that time, Bessel suspected a "dark" companion star and lo and behold, with better telescopes, Alvan Graham Clark was able to rob this "dark" star of its darkness almost 20 years later. It is not surprising that many astrophysicists initially assumed that observations in the outer regions of galaxies were incomplete. With better instruments and new windows in the electromagnetic spectrum, dark matter would reveal itself. With the advent of such a new window (i.e. X-ray astronomy at the beginning of the 1970s) the problem of missing masses did not vanish. It became quite the opposite: as you will read.

4.2 An Excursion Into X-ray Astronomy

With the advent of space flight, observations outside the Earth's atmosphere had become possible. The first rockets carrying X-ray detectors were now able to search the cosmos for sources of this high-energy radiation, at least for a few minutes. It is important to note that the Earth's atmosphere protects us from harmful X-rays from the cosmos, a highly desirable property. More than 70 years after the discovery of X-rays in the laboratories of physicists, it was finally possible to use rocket technology to escape the protective shell of the atmosphere and study the X-rays of the universe. Figure 4.6

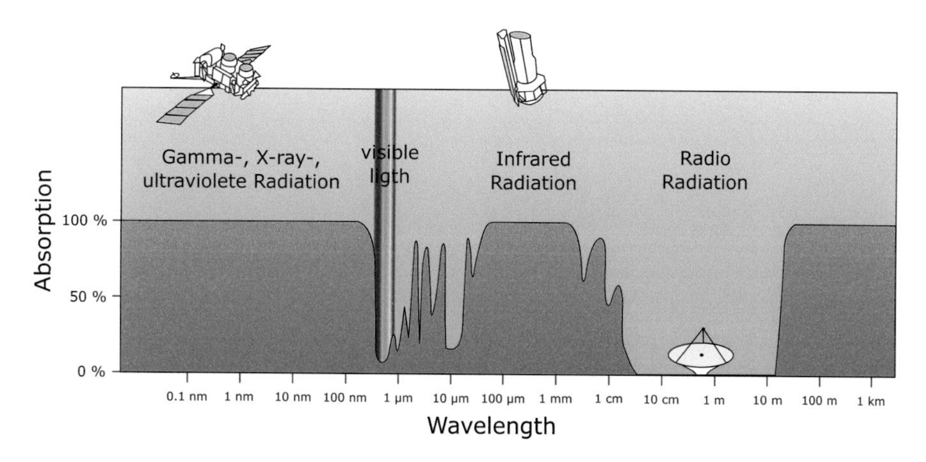

Fig. 4.6 The absorption of the Earth's atmosphere for different wavelengths of electromagnetic radiation. Source: By NASA (original); SVG by Mysid. [Public domain], via Wikimedia Commons

shows that our atmosphere does not allow observatories for X-ray astronomy on the ground.

The first source of X-ray radiation outside our solar system to be investigated was an object in the stellar constellation of Scorpio, which can be partially observed in summer in the visible night sky, even from Central Europe. This discovery was made in 1962 by a team led by Riccardo Giacconi (born in 1931) using a detector that reached a certain altitude with a rocket of the type Aerobee Hi Missile (see Fig. 4.7), an altitude at which X-rays from space were no longer absorbed by the terrestrial atmosphere. In these early days of unmanned space flight, it was not yet possible to operate a satellite in a stable orbit around the Earth. Therefore, an X-ray mission only lasted as long as the rocket was at an acceptable altitude, above 80 km. For the abovementioned

Fig. 4.7 An Aerobee Hi Missile. With this type, Giacconi and his team were able to discover an X-ray source outside our solar system for the first time in 1962. These rockets advanced to altitudes of up to 270 km and carried a payload of up to 68 kg during their flight. Source: *White Sands Missile Range Museum*

extrasolar first discovery, this duration was just 350 s. Today we know that Giacconi and his colleagues observed a binary star system that emits X-ray light and is located at a distance of about 9,000 Ly from us, that is, within our Galaxy. For this and his further work in the field of X-ray astronomy, Giacconi was awarded the Nobel Prize in Physics in 2002, together with the Japanese physicist Masatoshi Koshiba (born in 1926) and the American physicist and chemist Raymond Davis Jr. (1914–2006). During the following years, more and more X-ray sources in space were detected with such rocket experiments. All the way to extragalactic objects, which were first clearly detected in 1966 in the direction of M87, a large elliptical galaxy in the Virgo cluster of galaxies (Byram, Chubb & Friedmann).

At the beginning of the 1970s, three X-ray sources were known in the direction of the Coma, Perseus and Virgo galaxy cluster. In 1972, the first satellite *Uhuru* (see Fig. 4.8), which was specifically designed for X-ray astronomy was launched into a stable orbit. Its task was to produce the first map of the sky in the X-ray range of the electromagnetic spectrum - over a period of 3 years. The early observational results of the then new discipline called X-ray astronomy showed that many galaxy clusters are sources of X-ray radiation. One discovery was astonishing in particular: The X-ray sources were diffuse and emitted over the entire galaxy cluster. Assuming that the diffuse X-ray source was at the same distance as the galaxy cluster, the X-ray radiation summed up as energetic as the X-ray radiation from 100 billion suns. And this X-ray radiation did not change over the observation periods, it remained completely constant.

Uhuru was followed by several X-ray satellite missions, all of which reproduced the previous observations and substantiated them in even greater detail. The current main instruments of X-ray astronomy are *Chandra* and *XMM Newton*. Figure 4.9 shows an observation composite of a Chandra X-ray observation and an optical observation of the European Southern Observatory of the galaxy cluster Abell 383. The diffuse X-ray radiation (colored in violet) between the galaxies can be seen very clearly. In the 1960s, no such high-resolution X-ray observations were available, and the question of the origin of this radiation was far from clear. Nor was it known at all what physical process could emit so much X-ray radiation in such a large volume of space. First of all, it had to be clarified whether individual X-ray sources in the galaxies were the origin of this radiation or whether a new medium between the galaxies - the so-called *intergalactic medium* - was responsible for this. As early as 1966, still based on scarce measurement data, attempts were made to approach this complex of questions. In a work by James Felten et al. (1966), some answers that were still highly speculative

Fig. 4.8 Marjorie Townsend and Bruno Rossi at the NASA Goddard Space Flight Center with the first satellite Uhuru, designed solely for X-ray observations. The name Uhuru means freedom in Swahili. The name is to remind us of the launch off the coast of Kenya. Source: *By NASA (Great Images in NASA Description) [Public domain or Public domain]*, via *Wikimedia Commons*

at that time can be found, but they turned out to be surprisingly robust over time. The authors argued as follows:

If it is true that there is an enormous proportion of dark matter in the Coma cluster of galaxies, then one can ask what properties this mass would have to have in order to emit the observed X-rays. The hypothesis that all dark matter consists of the most abundant element in the universe, hydrogen, has been put forward. Thus, if the entire missing mass were evenly distributed over the volume of the Coma cluster of galaxies from which the X-rays could be observed, the X-rays would be theoretically explainable, but only if the hydrogen gas between the galaxies had a temperature in the range of 200 million Kelvin. In this case, the X-ray observations could be explained by a well-known process in physics, *thermal Bremsstrahlung*.

Fig. 4.9 X-ray (violet) and optical observation of the galaxy cluster Abell 383. The diffuse X-ray radiation between the galaxies is clearly visible. Source: *X-ray: NASA/CXC/Caltech/A.Newman* et al.*/Tel Aviv/A.Morandi & M.Limousin; Optical: NASA/STScI, ESO/VLT, SDSS*

At temperatures of 200 million Kelvin, hydrogen occurs only completely ionized. This means that only free charge-carriers (i.e. electrons and protons) that are no longer bound in neutral atoms are present. The existing charge fields of these protons and electrons cause permanent attraction and repulsion between them. This means permanent accelerations of charge carriers. And whenever charged particles are accelerated, they emit electromagnetic radiation, an effect known as Bremsstrahlung (braking corresponds to *negative* acceleration), see Fig. 4.10. With the help of the theory of the origin of Bremsstrahlung in ionized gases, it is possible to calculate exactly what temperature and density the gas between the galaxies would have to have in order to explain the observed X-rays of the Coma cluster of galaxies.

The result was and is astonishing: The plasma must be several million degrees Celsius hot and at the same time extremely thinly distributed - for example, one free charge carrier in a cube with an edge length of 10 cm. For comparison, dry air at 20 °C at sea level contains 2.5×10^{22} particles in the same volume. In other words, this extremely low density is comparable to the mass of two sugar cubes homogeneously distributed in a volume the size of Jupiter, the largest planet in our solar system. Our Earth fits about 1,400

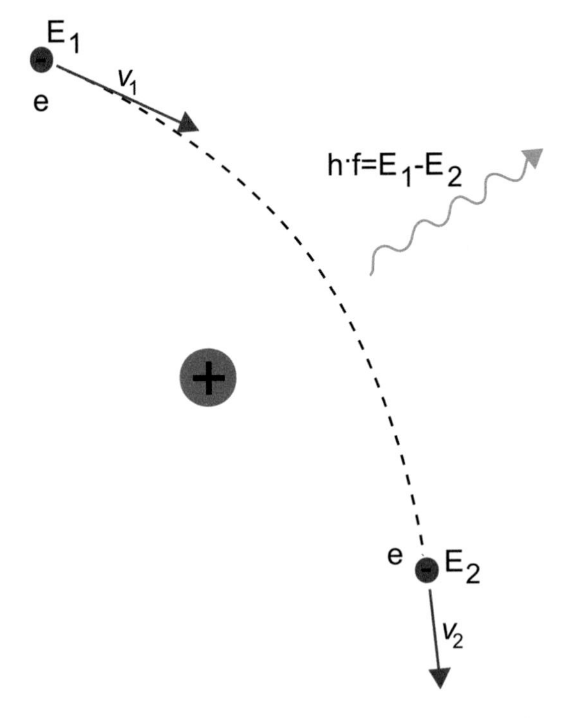

Fig. 4.10 Thermal Bremsstrahlung (sometimes also called free-free radiation). The acceleration of a charged particle (e), such as an electron in this case, caused by interaction with a positive charge (here that of a proton), produces electromagnetic radiation with an energy proportional to the acceleration of the charge carriers

times into this volume. One can also detect spectral signatures of elements such as iron or magnesium in this plasma.

This bold assumption by Felten et al. was supported by the velocities of the galaxies in the Coma Cluster measured from the galaxy spectra. If the particles of the gas had a comparable mean velocity to that of the observed galaxies, the mean temperature could also be explained exactly in the range of 70 million Kelvin. Therefore, in 1966, a possible candidate for dark matter in galaxy clusters had come into the light of scientific research. At the time of Zwicky, no one would have suspected that such a *medium* (usually referred to in literature as *intracluster medium*) could exist. Once again, an old rule has proven true: If you open a new observation window into space, you are almost always surprised by something unimaginable.

In order to put the model of thermal Bremsstrahlung of a very hot but extremely thin gas between the galaxies on a better foundation, more precise spectroscopic investigations of the X-rays were needed. Almost 10 years after the launch of Uhuru, a new generation of X-ray telescopes succeeded in doing so. An X-ray satellite called *Einstein* was able to record far better spectra of the X-rays from galaxy clusters, which could then be compared with theoretical spectra of various X-ray emission mechanisms. And lo and behold, Felten and his colleagues should be right. The theoretically determined spectra of thermal Bremsstrahlung could be reproduced excellently by the observed spectra.

Since the turn of the millennium, two additional X-ray satellites have been in operation: on the one hand, the NASA-operated *Chandra X-ray Observatory* and on the other hand the *XMM Newton* of the European Space Agency (ESA). And since 2005, there is another X-ray telescope of the Japanese space agency JAXA called *Suzaku*. These technically very complex instruments made it possible to study the properties of the X-ray emitting gas within the galaxy clusters much more precisely. It was now possible to determine the composition of the gas spectroscopically and to map the mass and temperature distribution of the intracluster medium in detail. But for the moment one important question remains unanswered: How much mass is actually bound in this hot gas? Can it fully explain dark matter?

The Intracluster Medium - Radiant Indicator of Dark Matter

It was Susan Lea et al. who first systematically investigated this question in 1973. Their approach was similar to that of Fritz Zwicky when he used the radial velocities of individual galaxies to determine the total mass of the Coma Cluster. The working hypothesis was that all the intracluster medium and all the galaxies in a galaxy cluster would have to follow a specific spatial distribution if they were in a state of equilibrium; a state in which the dynamics of the galaxy cluster can be described by the virial theorem and in which there are no major changes in the density distribution of the galaxies and the intracluster medium. In other words, if one knows the density distribution of the intracluster medium, one can draw conclusions about the total mass present. Since the intensity of the X-rays increases with the square of the density of the emitting medium and one can neglect internal radiation absorption in the very thin intracluster medium, the X-ray observation of the intracluster medium directly reflects the density distribution. However, an assumption about the distribution of the galaxies and the gas in all three directions in space has to be made, as unfortunately, only a two-dimensional projection can be observed in the sky. This was exactly the approach of Susan Lea and her colleagues. In the early 1970s, they took data from the X-ray satellite *Einstein* from the Perseus, Coma and Virgo clusters of galaxies and observed the X-ray radiation intensity on concentric rings of equal area around the respective X-ray maxima. The result for the Perseus cluster of galaxies is shown in Fig. 4.11. The observed values are approximated with a theoretical density distribution (solid line). Such a profile is also called King profile, after the astrophysicist Ivan King, who found this profile in the brightness distribution of globular clusters.

By assuming a spherically symmetrical distribution of the intracluster medium from the intensity of the X-rays in the ring-like regions around the center of the emission, one can now calculate the density of the gas per spherical shell. An important assumption here is a uniform temperature of the X-ray gas. Based on observations, Lea and colleagues assumed 1×10^8 K. In addition, a gas mixture of 90% hydrogen and 10% helium was used, a theoretically well motivated value.

The results were astonishing. The model for the total mass of the X-ray emitting gas of the Perseus galaxy cluster was $4(\pm 2) \times 10^{14}$ solar masses. The total mass of the cluster, derived from the dynamics of the galaxies in

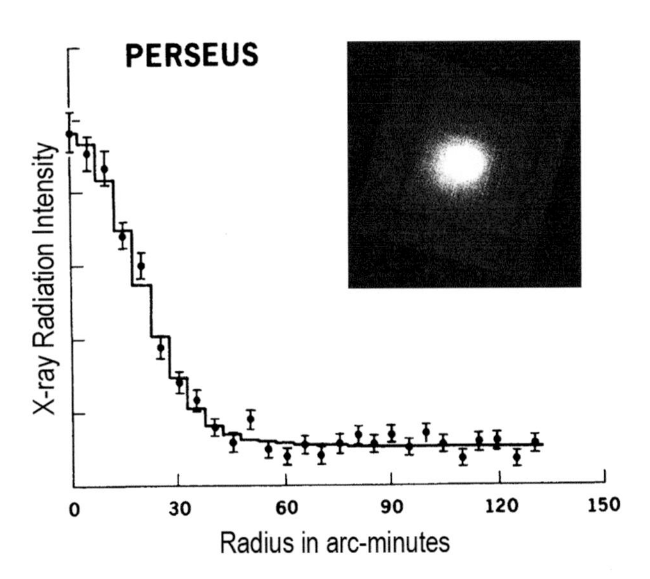

Fig. 4.11 Distribution of X-ray radiation on concentric rings of equal area around the maximum of the measured X-ray emission of the Perseus galaxy cluster. The observed data are approximated with a theoretical density distribution (solid line). The maximum X-ray radiation intensity is found at 0 in the center of the galaxy cluster. The radiation intensity decreases toward the outside. Source: Lea et al (1973) and *Nasa - Einstein Observatory Mission*

the Perseus galaxy cluster and determined by the velocity of the galaxies, is 2×10^{15} solar masses. The mass present in the galaxies of the Perseus cluster corresponded to 1×10^{14} solar masses. What a result! There is much more mass in the intracluster medium than in the galaxies of the Perseus cluster, but still a factor of 10 less than necessary to explain all the dark matter. The picture is the same for all the clusters studied in this paper. The gap could not be closed. Today, after more than 40 years and countless observations of other galaxy clusters with ever better instruments, the intracluster medium can be assigned the following properties: The main process behind the X-ray emission is the thermal Bremsstrahlung. It is impossible that the intracluster medium embodies the hoped-for dark matter in galaxy clusters that was once so elegantly discovered by Zwicky. The mass budget in galaxy clusters can be represented as follows: About 5% of matter is present within the galaxies, about three times more mass is in the hot gas between the galaxies, the intracluster medium, and the rest of 80% remains in the dark, see Fig. 4.12. Actually, the hot X-ray gas has exacerbated the problem of dark matter. Especially the high temperature of the intracluster medium indicates the presence of an enormous mass accumulation within the galaxy clusters, which holds the gas together and heats it up to these temperatures.

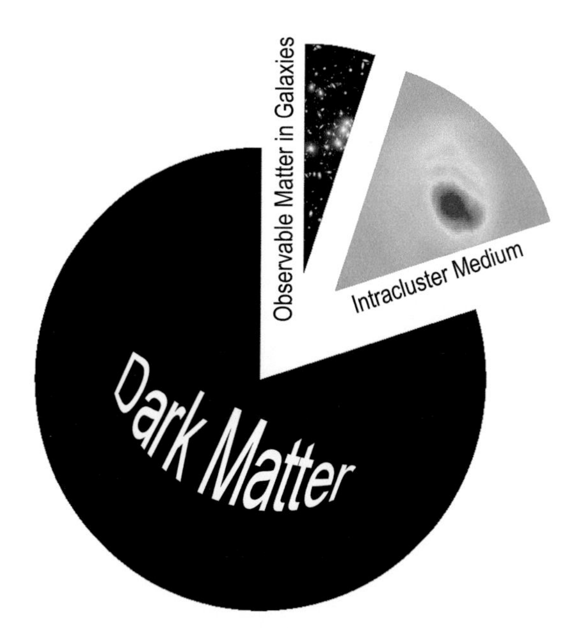

Fig. 4.12 The typical mass budget in galaxy clusters: About 5% of matter is observable within galaxies. Three times more mass is bound in the hot gas between the galaxies. About 80% of the total mass is still unknown in its nature - Dark matter

But more about this later, when you will learn more about structure formation in the universe.

4.3 The Gravitational Lensing Effect

In addition to the hot X-ray emitting gas between the galaxies, there is another important indication of the presence of a very massive dark component of the universe. These are strongly distorted images of astronomical objects; images whose distortion we can understand with the help of Albert Einstein's general theory of relativity and which also tell us a lot about the distribution of mass in the universe.

The central idea of Albert Einstein's general theory of relativity is: Mass changes the geometry of spacetime - or in short mass *bends* spacetime. It is observed that light follows the fastest path through spacetime, which in Newton's world view corresponds to straight paths. If a high mass concentration bends spacetime, curvilinear paths of light appear, and the higher the mass concentration, the stronger the curvature of spacetime. The idea of deflecting light by mass concentrations was first published in 1804 by Johann Georg von Soldner (1776–1833), a German geographer and astronomer. Soldner's consideration was that light consists of massive *light particles* with masses and would therefore have to be deflected by a gravitational field, such as our Sun. He calculated the resulting deviation for our central star, our Moon and the Earth. He concluded, for example, for the Sun a value of angular seconds, an immeasurably small value for those times. Thus, Soldner writes almost apologetically:

> By the way, I don't think I need to apologize for publishing this paper, since the result is that all perturbations are imperceptible. For we must be almost as interested in knowing what is there according to theory but has no noticeable influence on practice as we are interested in what really has influence in respect of practice. Our insights are equally extended by both.
>
> Source: "On the deflection of a light ray from its rectilinear motion, by the attraction of a celestial body at which it nearly passes by" ("Über die Ablenkung eines Lichtstrals von seiner geradlinigen Bewegung, durch die Attraktion eines Weltkörpers, an dem er nahe vorbei geht"). *Astronomical Yearbook for the year 1804* (*Astronomisches Jahrbuch für das Jahr 1804*).

Figure 4.13 shows an original sketch from Soldner's work. One recognizes the deflection ω of a light ray between the two lines DA (deflected by the

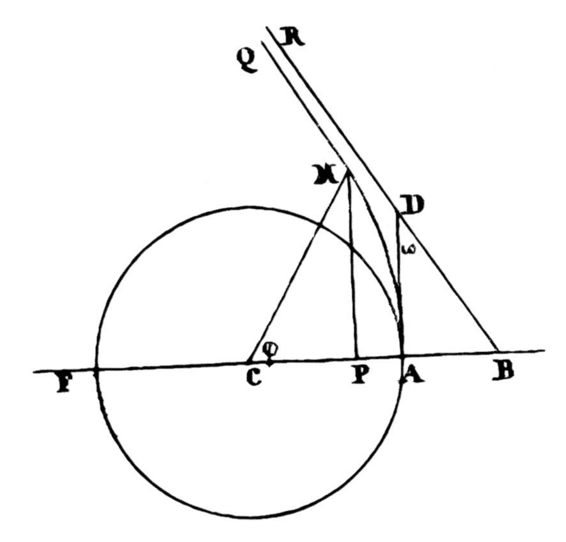

Fig. 4.13 Original illustration from Soldner's work "On the deflection of a light ray from its rectilinear motion, by the attraction of a celestial body at which it nearly passes by." *Astronomical Yearbook for the year 1804*. The deflection angle ω is drawn between the lines DA and DB

Sun) and DB (without influence of the Sun). Soldner describes the deflection in his approach as dependent on the speed of light and the mass of the deflecting body.

Light deflection is motivated completely differently in general relativity than in Soldner's theory. It turns out that Soldner's model for the interaction of mass and light particles (caused by a gravitational force) cannot capture the observations from today's point of view. In Newton's world view, the dynamics (motion) of bodies is described by a force acting between them. In the model of general relativity, on the other hand, an interaction of this kind does not exist; here the concept of the curvature of spacetime takes its place. Light follows the shortest path through space and time. The calculated angle of deflection by the Sun in general relativity is $1.75''$, exactly twice the value Soldner had published. By measuring the positions of the stars in the vicinity of the Sun, the predicted value of the general relativity could be confirmed. This was first achieved on May 19, 1919 during a solar eclipse on the volcanic island of Príncipe off the coast of West Africa by Sir Arthur Eddington, one of the first experimental confirmations of Albert Einstein's theory.

Whenever an established theory is superseded by a more general one, it is necessary to examine whether the correct predictions of the old theory still apply in the new model. Are the correct results of the old theory a subset of the new model? The correspondence principle applies: the correct predictions of the old theory must be reflected in the new one - a kind of theoretical Matryoshka.

For the investigation of dark matter within the framework of general relativity, the following therefore applies: If there were differences in the results of the mass determinations of objects obtained Newtonian mechanics and general relativity, which in principle could be described exactly by both theories, then the correspondence principle would be violated. So there would be another model, not yet developed, that would have to combine both theories. In the field of large-scale structures in the universe such as galaxy clusters or galaxies, the investigation of light deflection is another test of general relativity.

Figure 4.14 shows schematically the gravitational lensing effect caused by the mass distribution in galaxy clusters.

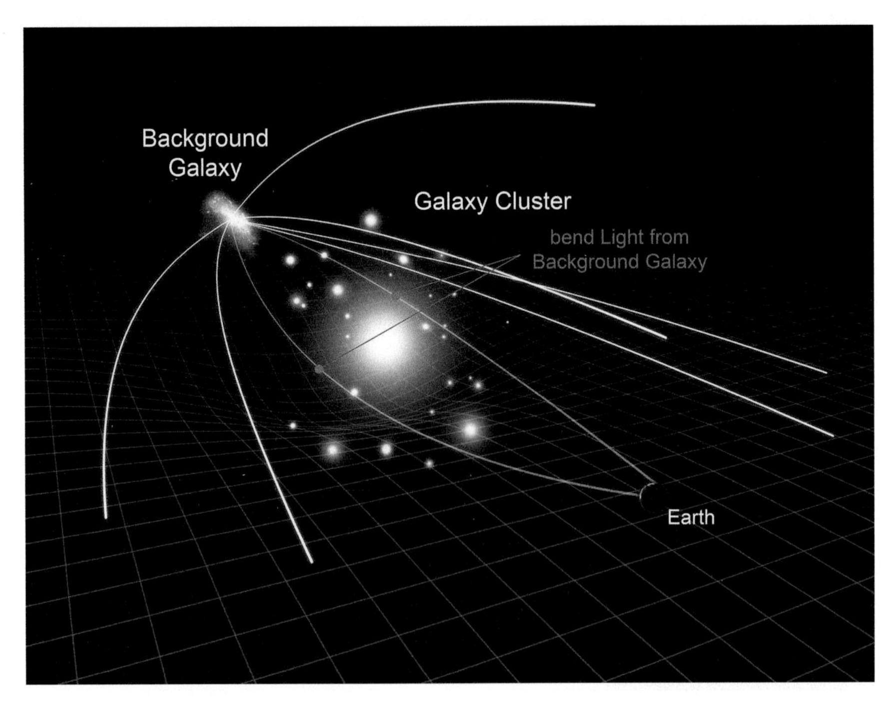

Fig. 4.14 Gravitational lensing effect caused by the mass of a galaxy cluster. The image of a background galaxy is distorted by the curvature of spacetime caused by the foreground mass - the galaxy cluster. The result is a distorted image of the background galaxy at the position of the observer on Earth. Source: *NASA, ESA & L. Calçada*

A galaxy cluster bends spacetime at his location by its large mass. Light rays from a galaxy behind us are deflected on their way through the bend spacetime and arrive distorted at the observer on Earth. Since the spacetime curvature resembles a gigantic converging lens in its effect, more light rays can be directed to the observer under certain circumstances. The result is then an increased brightness of the background galaxy. This is a fortunate circumstance that allows us to study distant objects that would remain hidden in the darkness of the cosmos without a gravitational lens. Figure 4.15 shows the observation of a distorted galaxy image caused by a galaxy cluster in the foreground. This image was taken with the Hubble Space Telescope and clearly shows us how strongly the enormous masses of the galaxy cluster bends spacetime. However, this effect also occurs with smaller masses, such as black holes or other massive, compact objects that can act as gravitational lenses.

The distorted background galaxy is almost 10 GLy away, whereas the galaxy cluster, which bends spacetime at his location very complexly due to its mass-distribution, is located at a distance of 5 GLy.

Fig. 4.15 Gravitational lensing effect observed with the Hubble Space Telescope in the galaxy cluster RCS2 032,727–13,262: A background galaxy is seen in a strongly distorted bluish color. Source: *NASA, ESA, J. Rigby (NASA Goddard Space Flight Center), K. Sharon (Kavli Institute for Cosmological Physics, University of Chicago), and M. Gladders and E. Wuyts (University of Chicago)*

The distortion of the image depends primarily on the mass and its distribution in the gravitational lens. If the gravitational lens were a point mass, the resulting deflection angle would be quite easy to calculate. The big problem with using the gravitational lensing effect as a mass estimate for galaxy clusters is that a galaxy cluster is not a point mass. In order to determine the mass distribution in a galaxy cluster, mass distribution models must be assumed - a certain uncertainty factor. Nevertheless, it has been shown that the masses of galaxy clusters, which were determined on the one hand by the gravitational lensing effect and on the other hand by the virial theorem, agree well. And again, dark matter is needed to bring the model predictions of general relativity into agreement with the observations.

Summary

While in the first half of the twentieth century, it was a few astrophysicists who drew attention to the problem of missing mass, the second half of the last century was marked by an explosion of evidence from observers. From the dynamics of galaxies, the intracluster medium, the hot gas observed in the X-ray regime of the electromagnetic spectrum, to the distorted images of distant objects by clusters of galaxies as gravitational lenses: Everywhere on large scales in the universe, a massive discrepancy between observed and gravitationally determined mass was detected. However, the concept of dark matter paved the way to a deep understanding of the origin of the structures in the universe. You will read in the next chapter that dark matter, along with dark energy, is the key component in the Standard Model of contemporary cosmology, and that it allows us to elegantly describe the formation of all the galaxies and clusters of galaxies, and even the cosmic web itself.

References

Babcock HW (1939) The rotation of the Andromeda Nebula. LOB 498:41–51
Brown H, Hazard C (1951) Radio emission from the Andromeda nebula. MNRAS 111
Felten J et al (1966) X-Rays from the coma cluster of galaxies. ApJ 146:955–958
Lea S et al (1973) Thermal-Bremsstrahlung Interpretation of Cluster X-Ray Sources ApJ 184:105–111
Rubin V, Ford K (1970) ApJ 159

Van de Hulst HC, Raimond E, van Woerden H (1957) Rotation and density distribution of the Andromeda nebula derived from observations of the 21-cm line Bull. Ast. Inst. Neth. 14,1 oder Volders L. 1957. Bull Ast Inst Neth 14:323

5

How Dark Matter Dominates our Model of Structure Formation in the Universe

As you have already read, there are many observations that suggest a dominant dark mass component in the universe. Through these observations and their theoretical interpretation, the dark matter model has become increasingly manifest in astrophysics. Modern cosmology and the model of structure formation embedded in it represents another important step towards general acceptance of the phenomenon dark matter. The progress in observational cosmology and its theoretical modeling in recent decades has led to an unexpectedly precise model of the formation and evolution of cosmological structures such as galaxies and galaxy clusters. Some astrophysicists currently speak of an "exact" cosmological world model, which only needs to be fine-tuned in its parameters by means of ever better observations.

In this chapter, these developments are presented in more detail and the central role of dark matter is elaborated. This is because the success of the standard model of modern cosmology, which is predominantly based on "dark" components, has established the phenomenon of *dark matter* and *dark energy* in current natural science. But let us first address the central key observations of cosmology, the measurements and their interpretations that leads to a dynamic, expanding universe; a universe that is said to have originated in a Big Bang.

We have already come across one central observation. It was made by Edwin Hubble in the 1920s when he studied the redshift of extragalactic objects and discovered that there is a strong correlation between distance and redshift of the spectra of the light emitted by distant objects: the further away, the greater the redshift in the light emitted by the objects. If we

W. Kapferer, *The Mystery of Dark Matter*, Astronomers' Universe,
https://doi.org/10.1007/978-3-662-62202-5_5

interpret redshift as the relative speed of the radiation source to the observer, we can conclude that the further away, the faster the objects move away from us. The interpretation of this observation by a singular event, the Big Bang, is a cornerstone of the current cosmological model. But there is one central observation that is another pillar: cosmic microwave radiation. An observed small difference in the temperature distribution of this radiation with maximum effect for all of us.

5.1 Small Difference with Maximum Consequences - Cosmic Background Radiation

In many textbooks, it is often briefly and succinctly stated that the cosmic background radiation was discovered by chance in 1964 by Arno Penzias and Robert Woodrow Wilson. This is only partially true because it took an interpretation of this accidental discovery, which is often referred to as the "echo of the Big Bang." And this interpretation shows very well how theory and observation can work together. Remember, an observation or a model by itself never has the explosive power of a complementary duet.

More than 10 years after Albert Einstein presented his general theory of relativity in 1915, a theory that elegantly links matter, space and time, Abbé Georges Edouard Lemaître (1894–1966) published an exact solution of the equations on which Einstein based his theory. Lemaître, a Belgian priest and astrophysicist, showed that Einstein's fundamental theory theoretically allowed an expanding universe as a possible solution. A universe that evolves from a singularity and continues to expand over time. Initially, the idea of an expanding universe was not accepted by Einstein. He advocated the theory of a universe that is unchangeable on cosmic scales. But Hubble's observations and Lemaître's calculations were able to slowly change this view of the world. So Lemaître writes in his important publication of 1931 ("A Homogeneous Universe of Constant Mass and Increasing Radius accounting for the Radial Velocity of Extra-galactic Nebulae," MNRAS 91):

> … It remains to find the cause of the expansion of the universe. We have seen that the pressure of radiation does work during the expansion. This seems to suggest that the expansion has been set up by the radiation itself.

Is it possible to obtain information about this very early phase of the universe just after the Big Bang? This is an exciting question that has been tackled theoretically from a completely different subdiscipline of physics, nuclear physics.

It was the early, groundbreaking work by the Russian physicist George Anthony Gamow (1904–1968) and his team that theoretically described a background radiation of that Big Bang in the late 1940s - 16 years before its official discovery. Gamow and his colleagues wanted to establish a satisfactory theory for the origin of the chemical elements, a theory that envisaged a continuous build-up process of the chemical elements. The starting point for their considerations was a highly compressed neutron gas, which partially transformed into protons and electrons as the universe expanded and cooled. These building blocks could then form the first, lightest elements such as hydrogen. Within the framework of this model, Ralph Alpher (1921–2007) and Robert Herman (1914–1997), both American physicists and cosmologists, calculated in 1948 a "theoretically observable" background radiation of the early universe in the region of 5 K, only slightly above absolute zero. In summary, at the beginning of the 1950s, there was a complex theoretical picture regarding a possible background radiation in the universe - if you like, the theoretical picture of the echo of the Big Bang was already there.

The observability of this theoretically predicted background radiation was first published by the Russian astrophysicists and cosmologists Andrej Georgievich Doroshkevich and Igor Dmitriyevich Novikov in 1964, who even proposed an instrument to observe these remnants of radiation from the early phase of the universe, the 15-m horn antenna of the Bell Telephone Laboratories in Holmdel, New Jersey, USA. Both had knowledge of this horn antenna, see Fig. 5.1, based on a largely technical publication by Edward Ohm in 1961 about the attempt to use this antenna to measure radio signals reflected from balloons.

Once again, the insight of Sir Arthur Eddington is proving to be correct:

> ... I hope I shall not shock the experimental physicists too much if I add that it is also a good rule not to put overmuch confidence in the observational results that are put forward until they have been confirmed by theory.

Although some observations pointed to background radiation in the range of theoretically predicted frequencies, this was not published accordingly. There was no connection to a theory that exactly predicted this type of radiation. Only the two American physicists Arno Penzias and Robert Woodrow

Fig. 5.1 The 15-m horn antenna of the Bell Telephone Laboratories in Holmdel, New Jersey. It was originally built in 1959 to develop the communications technology for NASA's early satellite programs. Source: Great Images in NASA

Wilson took the important step of interpreting their observation as background radiation of a cosmic expansion. In the 1960s, both investigated reflected radio signals from high-flying balloons. During their investigations, they discovered a noise of microwave radiation that could be observed everywhere in the sky. After a thorough analysis of the measurements and a systematic investigation of all possible explanations, they came across a publication by Philip James Edwin Peebles, a Canadian astrophysicist and cosmologist, in which they wrote about theoretical background radiation from the early days of the cosmos in the microwave range. The frequency range was exactly the same as the one they observed and in which they could measure a signal across the entire sky. The observation had enormous significance because it exactly fitted the theoretical models and was thus robbed of its "insignificance." In 1965, Penzias and Wilson published the observation of the cosmic background radiation and, at the same time, Robert Henry Dicke (1916–1997) and co-workers, in whose group Peebles was in, published the theoretical model for understanding the observed signals.

What Exactly is Observed in the Cosmic Microwave Background?

The cosmic microwave background radiation is an electromagnetic radiation in the wavelength range of a few centimeters. No matter where you look in the sky, it can be observed in all directions. If you consider the permeability of the atmosphere to electromagnetic radiation, see Fig. 4.6, this type of radiation is precisely at the short-wave edge of radio astronomy, where the atmosphere becomes permeable to radio radiation. Penzias and Wilson discovered a signal of a frequency of 4,080 MHz, that is, the wavelength of about 7.3 cm; a signal that does not come from a specific direction, but can be observed almost everywhere in the sky in almost the same way. Measuring this signal on its own is one side of the coin, the theoretical work of Dicke et al. (1965) was just as important. Penzias and Wilson also recognized this and so it is not surprising that the theoretical work on the prediction of the signal was published in the same journal of the same issue exactly before the observation was published, a nice gesture. The content of the theoretical work was not only the theoretical prediction of a signal at that wavelength but also a model for the origin of the radiation. The work also gave information about the shape of the signal. The title of the paper was "Cosmic Black-Body Radiation" (Dicke et al. 1965).

What was the origin of this signal? The idea was based on the model of the Big Bang, which was well motivated by Hubble's observations. In it, the young universe was in a very hot and dense state. Matter and radiation existed in a state of equilibrium, which led to thermalization of this "fireball." In short, the very young universe had "almost" the same temperature everywhere. This temperature can be assigned a radiation distribution. The form of this pure temperature radiation is known in literature under the term blackbody radiation and was first described theoretically by Max Planck. We have already learned about it in the chapter *The art of weighing a star*. During the further expansion phase of the universe, there was a time when the temperature of the universe fell below a critical value. During this time, the many free protons and electrons could combine in pairs to form neutral hydrogen atoms. The universe became transparent, radiation and matter decoupled. From this point on, the shape of the spectrum of this radiation no longer changed. This means that the information about the physical properties of that young universe is now stored in this radiation. It is comparable to looking at the surface of a star. Due to the redshift of the spectra, this information was shifted to longer wavelengths, into the range of microwave radiation. In Fig. 5.2, the theoretical spectrum of the background radiation is

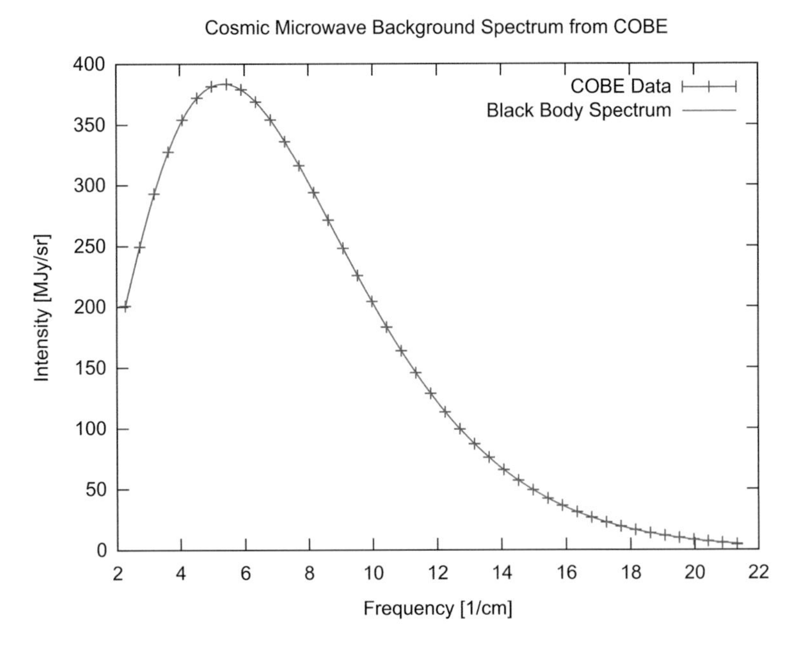

Fig. 5.2 The theoretical spectrum of the background radiation shown as a green continuous line. The red crosses show measurement results from the first satellite mission to explore the cosmic background radiation, called COBE (Cosmic Background Explorer, 1989–1993). Source: https://commons.wikimedia.org/wiki/File:Cmbr.svg

shown as a green continuous line. The red crosses show measurement results of the first satellite mission to study the cosmic background radiation, called COBE (Cosmic Background Explorer, 1989–1993). It is not very often that an astronomical measurement corresponds so exactly to a theoretical model.

The young universe was thermalized according to the standard model of cosmology. Matter and radiation were in energetic balance. When the universe became transparent to this thermal radiation, it could fill the expanding universe unhindered. Due to the cosmological redshift, this radiation was changed in such a way that today it corresponds to the radiation of a body "warm" by about 3 K. If you like, it was "cooled down" by the expansion. If one knows at what temperature the universe became transparent for this radiation, one immediately has the redshift and thus indirectly the information about the time when this happened. This transparency occurred when the universe was hot at about 3,000 K. This results in a time of about 380,000 years after the Big Bang. Since that time, the universe has expanded by a factor of about 1,000 according to the current parameters of cosmologists. As the universe continues to expand and thus this radiation is further redshifted, future observers will assign an ever lower temperature to this radiation until, in a hypothetical infinity, this echo in the vastness of space is silenced so that the universe continues to expand.

But now back to the exact measurement of this signal, which is so important for cosmologists. For this task, in addition to several balloon missions, there were three probes until 2014, the Cosmic Background Explorer - COBE (1989–1993), the Wilkinson Microwave Anisotropy Probe - WMA0050 (2001–2010) and the Planck probe (2009–2013). All of these probes had the task of producing an accurate temperature map of the entire sky in the region of the 3 K background radiation. This was achieved with ever increasing resolution, both spatially and spectrally. However, it could be seen quite early that there is a small temperature fluctuation in the cosmic background radiation around the precisely measured mean 2.72548 K in the range $\pm$ 0.00057 K. Figure 5.3 shows the sky map of the cosmic background radiation of the WMAP mission, and the recorded temperature fluctuations are in the range $\pm$ 0.0002 K around the mean temperature of the microwave background.

This map of the microwave background is a projection of the sky on a two-dimensional surface, a so-called Mollweide projection. This causes distortions at the poles; in Fig. 5.4, this projection is shown with the Earth's surface to make the distortions easier to see.

Fig. 5.3 A map of the entire sky in the range of the cosmic background radiation. The data are from NASA's WMAP. The recorded temperature fluctuations represent $\pm$ 0.0002 K. Source: NASA/WMAP Science Team WMAP # 121,238; 9-year WMAP image of background cosmic radiation (2012)

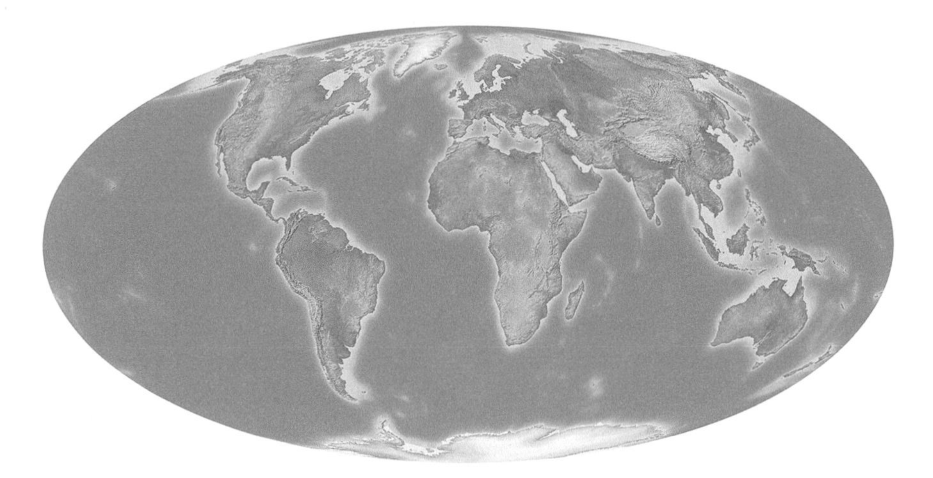

Fig. 5.4 World map created with Natural Earth data using the Mollweide projection, the same projection under which the cosmic background radiation is often displayed. Source: By Ktrinko (Own work) [CC BY-SA 3.0 (https://creativecommons.org/licenses/by-sa/3.0)], via Wikimedia Commons

Over the three missions, the spatial and spectral resolution of the instruments has been continuously improved, so that today much more detail can be seen in the temperature map of the early universe, see Fig. 5.5, important

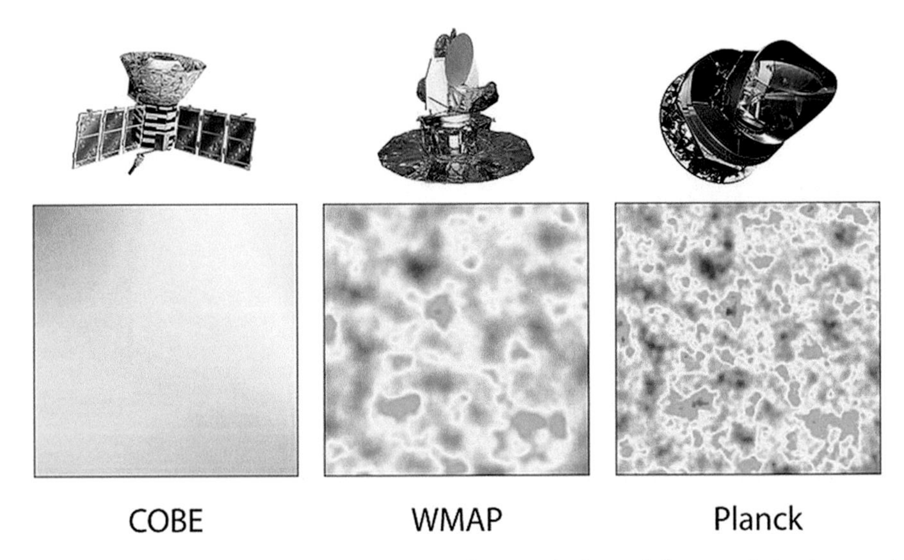

Fig. 5.5 The image shows the ever increasing resolution of the cosmic background radiation measurements over the three missions COBE, WMAP and PLANCK. In each box, a $10^\circ \times 10^\circ$ large section of the sky is shown. The temperature fluctuations are in the range of 0.0002 K. Courtesy NASA/JPL-Caltech

information that tells us a great deal about the origin of structures in the universe.

The key is to interpret the cause of the temperature fluctuations at the time of decoupling of matter and light. The hot and very dense "primordial soup" of photons, electrons, protons - known elementary particles - and dark matter, which is the source of cosmic background radiation, can be physically described as a very hot gas. Just as sound waves can propagate in the air, so do disturbances in this hot and dense gas at a much greater speed. The source of the "disturbances" in this primordial soup of matter is an interaction that inevitably occurs in the interaction of matter, that is, of masses: gravitation. Let us assume that there were areas where there was locally little more mass in this primordial mixture. This then led to particles being stronger attracted, resulting in a locally higher density. Whenever gas is compressed, the temperature of the gas increases accordingly. This results in an opposite effect: The high temperature causes it to expand again. However, this has the consequence that the particles expand again to a larger volume, which in turn leads to a cooling of the gas. The mixture becomes denser, has a higher mass concentration, gravitationally attracts more matter, becomes hotter and so on; a cycle that lasts until the universe has become transparent to radiation. At this point in time, the energy escapes into the vastness of space and the temperature and density associated with it is preserved for us observers. Of course, the photons of the background radiation experience a number of other influences on their path through the universe, but by appropriately taking numerous effects into account, it is possible to extract a density map of the universe that can be described with a few parameters in the Standard Model of Cosmology.

The data reduction of the measured background radiation is technically very complex and computationally intensive, but can be managed with today's large computer systems.

In addition, one can also make physical assumptions about the properties of dark and ordinary matter in this model and see what the temperature fluctuations would look like with a certain amount of dark matter. In general, dark matter in all models would result in different lumps than ordinary matter alone and thus change the temperature map significantly. Unlike gas, dark matter does not feel any thermodynamic pressure, so the density distribution of dark matter is different compared to ordinary matter. On the other hand, without dark matter, the density fluctuations of ordinary matter would be much smaller because the compression of the gas would be less strong without the presence of the gravitationally cohesive dark matter. These scenarios can be run through with model assumptions until you have a model that best describes the cosmic microwave background. Currently, the best models require dark matter, so the strength of the temperature fluctuations in the microwave background alone, together with the cosmologists' models, allows indirect conclusions to be drawn about the existence of dark matter.

As a further aspect, the map of the microwave background can be used to delineate areas that are causally related and that have never been in direct contact with each other. What does causally related mean? Before the decoupling of microwave radiation, density fluctuations in this gaseous matter-energy "cloud" can be interpreted as sound waves. These waves have a speed of propagation in this medium and therefore pass over a certain area during their propagation. As a result, events in the microwave background are spatially delimited and causally related. At the time of decoupling, these areas of influence in the microwave background can be measured because they do not change anymore, they are "frozen" in this map. Now one can calculate these influence areas and compare them with the observed ones, and thus one can also conclude the curvature of the universe. If the observed quantities and the calculated ones are in agreement, general relativity tells us that the universe is flat. This means that the matter in the universe, which acts as a gravitational lens overall, has not caused any image distortion. If the observed areas of influence were too small, then the universe would have a lower density; if they were too large, a higher density in the universe would be needed to understand the differences in size. Currently, the data indicate that the curvature of the universe is flat when viewed on the largest scales. No gravitational distortion can be observed in the microwave background.

The result of the WMAP data after 9 years of observation, data evaluation and interpretation, with regard to the amount of matter types, is: 83% dark matter and about 17% matter known to us (Bennett et al. 2013).

We summarize: About 380,000 years after the Big Bang, the universe was composed of light and matter in thermal equilibrium. The temperature fluctuations were in the range of a few hundred millionths of a degree and the density fluctuations derived from this were in the range of 0.1 ‰ around the density mean value. The universe was about 3,000 K hot and neutral hydrogen could be formed. This led almost abruptly to the universe becoming transparent to the heat radiation of this primordial form. This process is called decoupling of the microwave background radiation.

With this knowledge of the early universe, we are now trying to approach the question of how the planets, stars, galaxies and clusters of galaxies could have evolved from such a cosmological primordial soup. The next section attempts to outline the current main direction of the model of the formation of the observable structures in the cosmos. This is because there is an answer to the question of how some 13 billion years after the cosmic background radiation was decoupled, density fluctuations occurred that differ by more than 10 orders of magnitude from those that appeared 380,000 years after the Big Bang. You guessed it, it is called gravitation. The fact that dark matter will be of particular importance in this context is probably no longer surprising at this point.

5.2 Mapping with Light

To understand the formation of astrophysical objects, a model of the framework in which all the events take place, a model of the stage, is needed. Since Hubble discovered the redshift of galaxies, the static picture of the universe had begun to falter, and the universe on large scales became a highly dynamic environment.

It is interesting to note that at the beginning of the twentieth century, a theory of spacetime emerged, namely Einstein's general theory of relativity, which was able to describe the observations of the following decades so well. Had it been developed at the time of Newton, the theory would have been dismissed as a pipe dream. Einstein's theory links the geometry of spacetime with the distribution of masses and energies within it. At Newton's time, space and time were fixed and separate aspects of the world. Space was linear and separate from time. Time passed equally for all observers, according to the experiences of the natural scientists of those days. However, Newton's model always fails where large masses are located in small volumes because very strong accelerations occur in their vicinity, which can be described better, more precisely, by the model of spacetime curvature of general relativity. A classic example of this is the orbit of the innermost planet Mercury around the Sun. In Einstein's model, the Sun bends spacetime at the orbit of Mercury. This results in a deviation between the orbit calculated according to Newton's model and that calculated according to general relativity. Einstein's theory can accurately describe the orbit of Mercury.

To know the rate of expansion of the Universe using general relativity, we need to make assumptions about the average density and energy of the Universe. Once we have made these assumptions, the model tells us how fast and whether the universe is expanding or even collapsing. These are important parameters that directly influence the formation of structures. Figure 5.6 shows solutions of the general theory of relativity for some of these assumptions. The x-axis shows the time in billions of years, with the zero point today, and the y-axis shows the average distance between two points (as an example between two galaxies). The graph shows the evolution

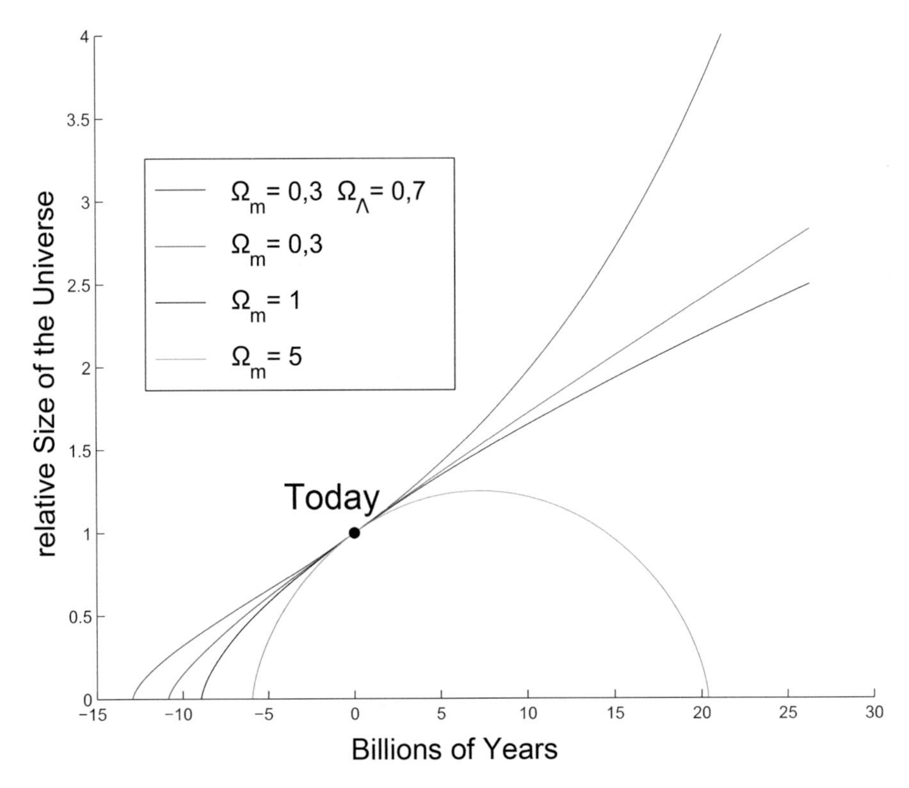

Fig. 5.6 The relative size of the universe for five different assumptions about the matter and energy content of the universe, where Ω_M stands for the density of matter and Ω_Λ stands for the energy density of the universe. The numerical values are given in the unit of critical density. The critical density is the density at which the universe stops expanding at infinity. If the universe has a higher matter density than 1 (e.g., $\Omega_M = 6$), the universe will collapse again

of the average distance between two distant points (e.g., galaxies in different clusters) in the universe for five different assumptions about the matter and energy content of the universe. These are designated by the letters Ω_M and Ω_Λ; whereby Ω_M stands for the density of matter and Ω_Λ stands for the energy density of the universe. The numerical values are given in the unit of a critical density. The critical density is the density at which the universe will stop expanding at infinity. If the universe has a higher density of matter (e.g., $\Omega_M = 6$), in the Standard Model of Cosmology, the universe will enter a contraction phase in the future and will collapse again in about 17 billion years. It is these two competing variables that determine what the fate of the universe will be, gravity and expansion. In the next section, you will learn about a competitor of gravity, dark energy. And you will see that, according to current models, gravity is not able to decelerate the expansion of the universe.

The interesting thing in Fig. 5.6, however, is the fact that by knowing the average distance at earlier times, one can derive the average energy and matter density of the universe and thus also make statements about the further development of the cosmos. By measuring distances in the cosmos, we can make well-founded statements about the density of matter and energy and the further development of the universe. But how can we measure these distances well? Let us remember the discovery of Hubble. We can do this by looking at the lines in the spectra of astrophysical objects and their displacement relative to the reference lines in a laboratory on Earth - keyword redshift.

Type Ia Supernovae and Dark Energy

The quest for which solution of Einstein's field equations, see Fig. 5.6, describes best the observations is a core task of cosmologists. The challenge is to measure distances to very distant objects in the universe as best as possible. These measured distances are then compared with the different models in Fig. 5.6. The model that comes closest to the measurements is accepted as the standard cosmological model.

To measure these long distances, we need an object that is very bright and of which we know how much light it emits - a very bright standard candle. In addition to the Cepheid stars already used by Hubble, there is another important class of astrophysical objects that is virtually made for this task - supernovae. From the observer's point of view, supernovae are stars that explosively increase their brightness. In the astronomical records of recent centuries, such new, bright stars have been observed on the firmament time and again. For example, in 1572, Tycho Brahe discovered a new star in the constellation Cassiopeia. In contrast to all other known fixed stars, however, their brightness decreases drastically over a relatively short period of time. Figure 5.7 shows such a supernova explosion in a distant galaxy. In the lower three boxes, a zoom on a galaxy is shown. You can see this galaxy with and without a supernova, and finally the image of the supernova, subtracted from the light of the host galaxy. You can see the enormous brightness of this astrophysical phenomenon. It is as bright as the whole galaxy.

For cosmological distance measurements, a very special class of supernovae is used, so-called Type Ia supernovae. The supernovae were categorized by spectroscopy. Type Ia supernovae do not show lines of hydrogen and helium. Little is known about the predecessor system of this class. Neither the

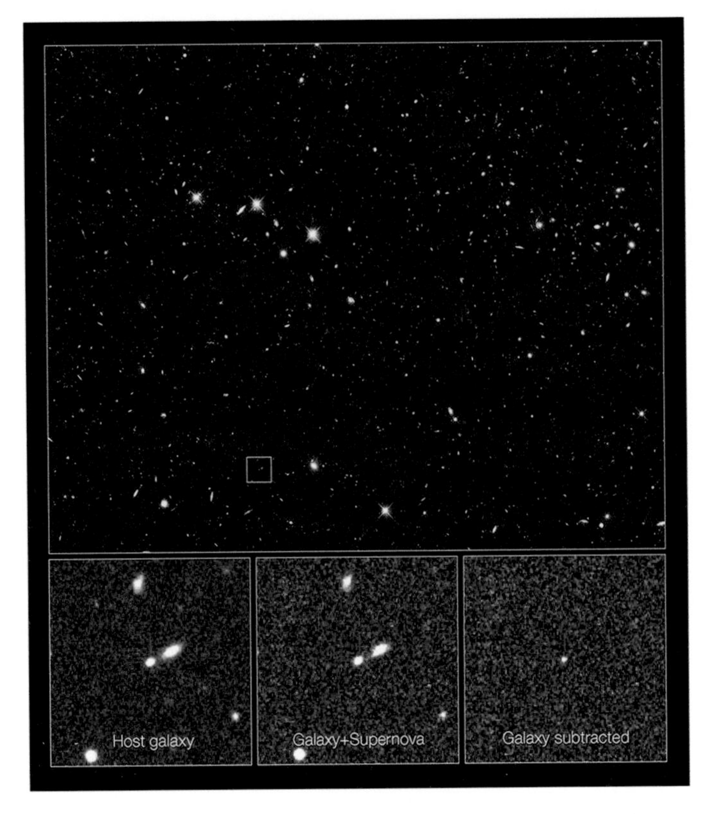

Fig. 5.7 A supernova explosion observed with the Hubble Space Telescope in a distant galaxy. In the lower three boxes, a zoom on the galaxy is shown with and without supernova. Finally, the image of the supernova, subtracted from the light of the host galaxy. You can see the enormous brightness of the supernova, which is almost as bright as the galaxy. Source: Cosmic Assembly Near-infrared Deep Extragalactic Legacy Survey SN UDS10Wil (CANDELS); NASA, ESA, A. Riess (STScI and JHU), and D. Jones and S. Rodney (JHU)

precursors nor all compact remnants of such an explosion have been observed in detail. Among the probable predecessors of such a supernova are binary star systems consisting of either a white dwarf and a central star or two white dwarfs. White dwarfs form a class of very compact stellar objects that are formed from rather light stars at the end of their lives. In both scenarios, the partners exchange matter due to their close orbits until the more compact partner gains enough mass to start a thermonuclear reaction. This happens at a rather fixed amount of accredited mass, which leads to a kind of standard explosion. We observe the light of this enormous explosion as a supernova. With the energy released in this process, the human energy demand of about

20 million billion earths could be satisfied for a billion years, truly astronomical dimensions (with an annual energy demand of 500 Exajoule).

Supernovae of this class show a very similar course of increasing and decreasing brightness, which can also be determined theoretically. This is shown schematically in Fig. 5.8. One can see the rapid increase and the typical decay of brightness over several months. Amazing is the maximum luminosity, which corresponds to several billion suns. It turns out that the maximum luminosity of Type Ia supernovae is strongly correlated with the course of the decay. If one can measure the brightness of such a supernova at different times, one can conclude its absolute brightness. Again, the same trick as with the Cepheid stars: If you know the absolute brightness, you can derive the distance to the supernova because this decreases with the square of the distance.

Because Type Ia supernovae are so luminous, these objects can be observed at great distances. They are exactly the kind of standard candles we need to define the current cosmological model. Figure 5.9 shows distances to observed supernovae of this type. The green dots show closer supernovae, the red dots very distant ones. One can see how important it is to evaluate the measurement data of the distant supernovae in the best possible way because it is not possible to differentiate between different models for closer supernovae. Only the distant supernovae show us that the model with the parameters $\Omega_M = 0.3$ and $\Omega_\Lambda = 0.7$ describes best the observations. Thus, we have a solid model of expansion for our theory of structure formation in the universe. It is important at this point to draw the reader's attention to the fact that an important assumption must be made here: The physics

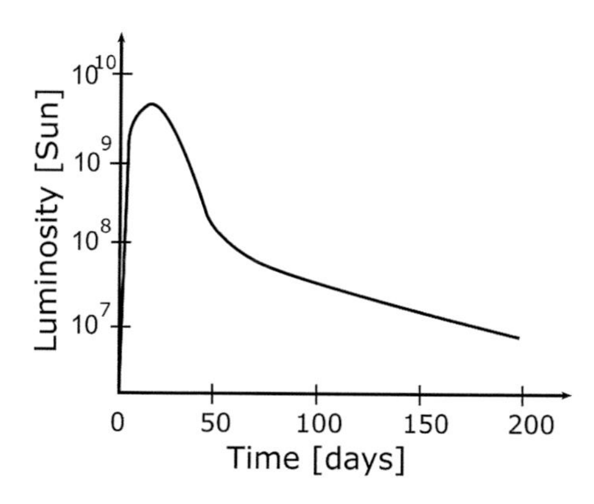

Fig. 5.8 Schematic supernova Ia light curve in units of solar luminosity. You can see the rapid increase and the typical decay of brightness over several months. The maximum corresponds to a brightness of several billion suns

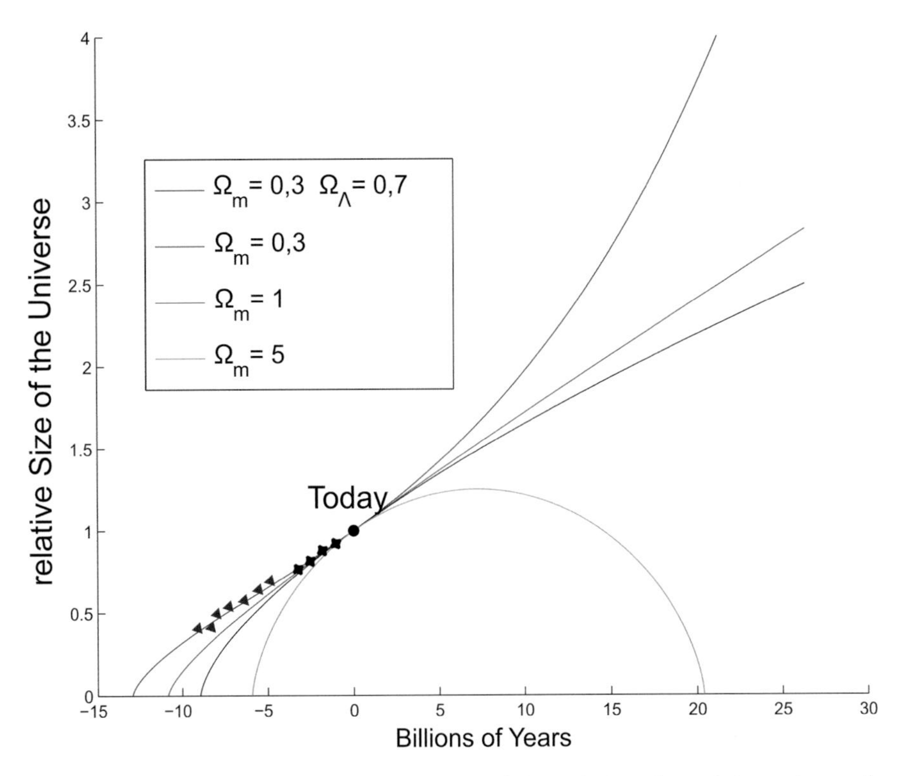

Fig. 5.9 The graph shows the development of the relative size of the universe for five different assumptions about the matter and energy content of the universe, where Ω_M stands for the density of matter and Ω_Λ stands for the energy density of the universe. The black star-like dots show closer, the dark red triangles, very distant supernovae. Thus, one can determine the parameters Ω_M and Ω_Λ

of supernovae are the same at all times. This almost sounds as if physicists were arrogant and imposed their reality on the universe. But let us assume for a moment that the laws of physics changed over time. How could one test assumptions in laboratories? How could we refute a hypothesis that the free fall experiments of Galileo Galilei a billion years ago would have had a different outcome than during his lifetime with current free fall experiments? Thus, such assumptions in the strict sense are primarily speculations that elude confrontation according to the rules of the natural sciences. This should not exclude the possibility that such speculations may be valuable and possibly enable new perspectives on problems. But let us return to supernovae. A strong evidence that the methods of physics can also be applied to cosmic events eight billion years ago is shown by the comparable brightness curves of the far-distant supernovae compared to the near ones.

You can see that the model with the density of matter $\Omega_M=0.3$ and the energy density $\Omega_\Lambda = 0.7$ describes best the measurements; in units of that critical density at which compensation for the expansion of the universe at infinity would occur. For this density, which is derived from the distances to the supernovae, it can again be concluded that all the directly observed matter in the galaxies, the intracluster medium and all the other astronomical objects is not sufficient to determine the mean density of matter in the universe. Dark matter is again needed to reach this density. To what extent the correspondence principle - every better model must reproduce the correct statements of a less far-reaching model - plays a role here with regard to dark matter is certainly an exciting, almost philosophical question. Another puzzle is the accelerated expansion of the universe derived from supernova observations. The model into which the observed supernovae fit best is an accelerated expanding universe. What exactly is the "engine" of this expansion remains a mystery. And as with the dark matter phenomenon, the term *dark energy* seems to describe this snapshot of scientific knowledge well. We know what properties these dark components possess, but we do not know their physical causation.

5.3 The Various "Body Mass Indices" of the Universe

Currently a model exists that describes the combined measurements of the microwave background and the cosmological supernovae distance measurements very well. The name of the model is Lambda CDM model (LCDM model), where CDM stands for Cold Dark Matter and Lambda for the accelerated expansion of the universe. Why dark matter is called cold here has to do with the hypothetical speed of dark matter particles and will be explained in more detail in the next chapter.

In medicine, people are often divided into health risk groups according to their body mass index (BMI). A single, easily derived quantity simply calculated from the mass of a person divided by the square of his or her length. Using this quantity, it is possible to statistically determine how the frequency of diseases is dependent on the BMI. Certainly, these studies represent models that should be critically questioned. But because BMI is easy to determine, the approach of giving statistical probabilities for the occurrence of diseases due to BMI is very interesting.

Table 5.1 Some current cosmological parameters of the Lambda CDM model of the Planck mission (September 2016). Source: Planck (2016)

Proportion of usual matter[a]	Ω_b	0.0486 ± 0.0010
Proportion of dark matter[a]	Ω_c	0.2589 ± 0.0057
Hubble constant	H0	67.74 ± 0.46 km $s^{-1}Mpc^{-1}$
Proportion of total matter[a]	Ω_m	0.3089 ± 0.0062
Proportion of dark energy[a]	Ω_Λ	0.6911 ± 0.0062
Critical density[a]	ρ_{crit}	$8.62 \pm 0.12 \times 10^{-27}$ kg/m^3
Redshift at the time of decoupling	z_*	1089.90 ± 0.23
Age of universe	t_0	13.799 ± 0.021 billion years
Age of universe at the time of decoupling	t_*	$377{,}700 \pm 3{,}200$ years

[a] In units of that critical density at which the expansion of the universe in infinity comes to a standstill

Well, cosmologists are no different; they also parameterize their models, which can then predict or reconstruct the future and also the past of the universe. These model predictions are then exact within their application limits and not filled with probabilities. However, the greatest uncertainty in the cosmologists' models is in determining the correct BMI of the universe. What is quite easy in humans turns out to be somewhat more difficult on cosmological scales. But let us first have a look at these parameters with their current values. One reason why we speak of precision cosmology today is the data on which cosmologists can build their theories. The parameters for their models are no longer subject to such large uncertainties as in earlier days. Let us consider the Hubble constant, which indicates the expansion rate of the universe. Until the mid-1990s, values for the Hubble constant in the range from 35 to 100 km/(s Mpc) were derived directly and indirectly from observations.[1] Today, this expansion rate is given with a fluctuation margin of $\pm 0.3\%$. Thus, the theorists are able to make equally accurate statements about the age, the different densities of matter and, last but not least, about the expansion behavior of the universe in the future with very precise parameters. Some results of the model based on the current state of observation are given in Table 5.1. These are the proportions of the different forms of matter in the matter density of the universe, the Hubble constant, the redshift at the time of decoupling of the background radiation and the age of the universe. As can be seen from the small values of uncertainty, the model already seems to be quite accurate. It is important to note at this

[1] https://www.cfa.harvard.edu/~dfabricant/huchra/hubble/

point that quantities are derived from observations with the help of models, leading to values as shown in Table 5.1. If the models are wrong, then the values are also wrong. You will see, however, that the model is capable of delivering very robust results that are comparable with the observation. Thus, it is currently the best description of the universe on large scales.

A Map of the Universe

Any model of structure formation on cosmological scales must be confronted by observations. The more detailed the observations, the better one can distinguish between different models and determine the parameters of a successful model. Additionally, one needs a detailed map of the universe in its closest vicinity and up to the limits of what can be observed. This is a challenge that astrophysics has successfully met for decades. Figure 5.10 shows the current deepest view in the optical region of the spectrum, the *Hubble eXtreme Deep Field (XDF)*. The observed area in the sky is about

Fig. 5.10 The currently deepest view in the universe in the optical part of the spectrum, the *Hubble eXtreme Deep Field (XDF)*. For size comparison, the image shows an area in the sky that is smaller than one hundredth of the area of the Moon in the sky. Source: NASA, ESA, G. Illingworth, D. Magee, and P. Oesch (University of California, Santa Cruz), R. Bouwens (Leiden University), and the HUDF09 Team

0.0000007% of the total sky. A very dark area of the sky was exposed for about 22.5 days. Almost all objects in this image are galaxies, only a few foreground stars can be seen in our galaxy. These can be recognized by the cross-shaped pattern that - caused by the design of the telescope - occurs with point-like sources such as stars. On the one hand, very faint galaxies can be detected, and on the other hand, very distant and correspondingly redshifted objects. The most distant galaxy is around 13 GLy away and was thus formed only 800 million years after the Big Bang. You can recognize the most different shapes and colors of the galaxies. If one wanted to observe the entire sky in this quality with the Hubble Space Telescope, almost 2 million years would be necessary. Therefore, in order to produce a map of the universe, we must use other instruments that can provide us with a less detailed picture of the universe, but which can map a larger section of the sky.

Such an instrument must measure the spectra of thousands upon thousands of galaxies in order to obtain a three-dimensional image of the distribution of galaxies in the universe thanks to redshift. Figure 5.11 shows such an instrument. The workhorse of modern "sky mapping," the *Sloan Digital Sky* Telescope in the mountains near Sunspot, New Mexico, USA. The telescope is operated by several institutes in the USA, Japan, Korea and Germany and was also co-financed by the Alfred P. Sloan Foundation, among others. Its main task is to measure the spectra of millions of galaxies and thus create a three-dimensional map of parts of the universe. In addition to the distribution, this also provides information about the physical properties of the objects.

A result of the sky survey with the Sloan Digital Sky Telescope is shown in Fig. 5.12. In this figure, we are in the center and each point represents a galaxy. The circle has a radius of about 2 GLy and the color indicates how old the stars in the galaxies are on average, with red representing an old and bluish representing a younger population of stars. The dark, wedge-shaped areas have been omitted because there our Milky Way disturbs the view too much. The figure shows a section through the three-dimensional data of the sky survey, a section that reveals one thing at a glance, the universe looks very similar in all directions. There are clusters of galaxies, the so-called galaxy clusters, and areas completely without galaxies, so-called voids. The figure resembles a web, which is why the term "cosmic web" is often used in this context. One also recognizes that there is no other kind of place in the universe. On large scales, every place in the universe is statistically similar and if you look around, there is no direction that is completely different to the observer. These two properties, on the one hand the homogeneity - the universe presents itself similar at all places - and the isotropy - there is no

Fig. 5.11 The workhorse of modern "sky mapping," the Sloan Digital Sky Telescope in the mountains near Sunspot, New Mexico. Source: SDSS Team, Fermilab Visual Media Services

direction in which the universe presents itself differently - have entered the literature under the term "cosmological principle."

Through ever deeper observations of the cosmos, structures on very large scales have been found again and again in recent years that question the principle of homogeneity and isotropy of the universe. A current one is the Hercules-Corona Borealis Great Wall, the most massive and largest structure in the universe so far, indirectly observed by high-energy gamma-ray bursts. It has a probable extension of about 10 GLy and consists of several clusters of galaxies and filaments of galaxies in between. Such a huge structure would represent a special, excellent location in the universe, which could put current models of structure formation in dire straits. However, only future observations will be able to clearly show whether or not it is a causally connected structure. Until then, the homogeneity and isotropy of the universe are well supported by many observations.

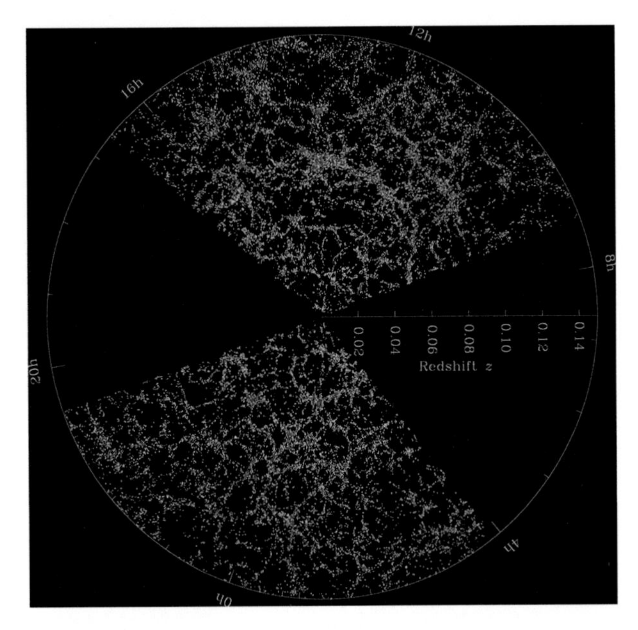

Fig. 5.12 Distribution of galaxies in a circle with a radius of 2 GLy. Section through a three-dimensional map created with the Sloan Digital Sky Telescope. The color scale indicates the average age of the galaxies, from red equal to old star population to blue equal to young star population. Source: Michael Blanton and the SDSS Collaboration, www.sdss.org

With such a map in hand, we can now set out in search of a model for the formation of structures in an expanding universe. The most successful model should predict not only the correct distribution, but also the statistical distribution of shapes and colors, an indicator of the age of the stars and galaxies in it. Truly no trivial undertaking.

5.4 A Universe in the Computer

A model of structure formation must primarily be able to describe the gravitational interaction of masses in an expanding space, since this is the dominant interaction on large scales. Unfortunately, there is no universally valid analytical solution to this problem. Even for more than two masses, there is no exact solution to describe all possible constellations and developments of the system elegantly by means of an equation. In celestial mechanics, systems with three masses have their own term, the three-body problem, which can only be solved exactly for special constellations and mass distributions.

To obtain a realistic model, however, many mass points are required, this is called a multi-body or *n-body* simulation. Numerical approximation methods are used, which can describe the temporal development of the positions and velocities of many gravitationally interacting bodies. Strictly speaking, these are approximations, analytically it is not possible to be exact, but the errors are controllable and - even more important - calculable.

A widely used algorithm for solving the multi-body problem is the Barnes-Hut algorithm named after Josh Barnes and Piet Hut, who published it in 1986 (Barnes and Hut 1986). But let us first look at the main part of any multi-body force calculation. Since the beginning of mechanics, the so-called superposition principle has been an integral part of force modeling. It states that the resulting force on a body is the sum of all individual forces acting on it. Let us now take a situation as shown in Fig. 5.13. There are bodies in a room, according to the distance and mass of the bodies, a force is exerted by one body on the middle one (shown here in red). The sum of all forces on the body here corresponds to the force marked in red. It is also called resulting force. Using Newtonian mechanics, we can assign an acceleration to this force because acceleration is force divided by mass. We can now do this for all the bodies in our example and we will know where the masses

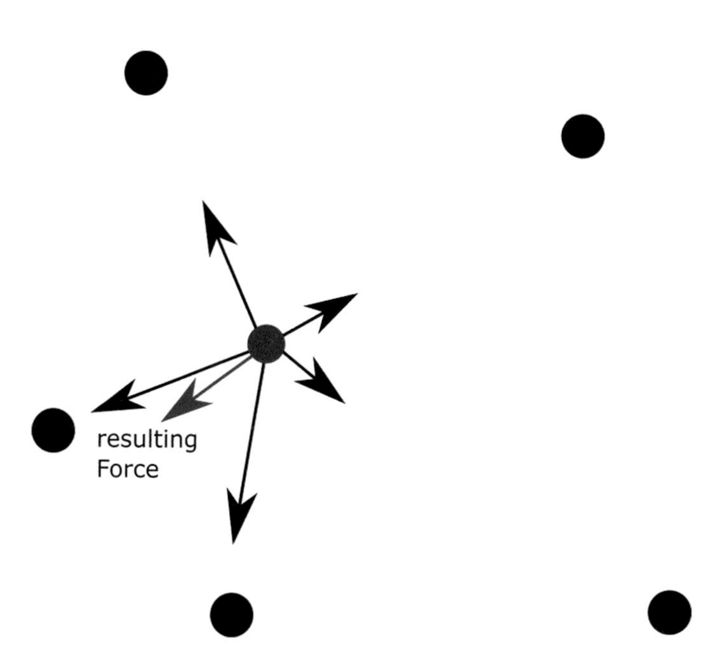

Fig. 5.13 The sum of all individual forces on a body climax in a resulting total force

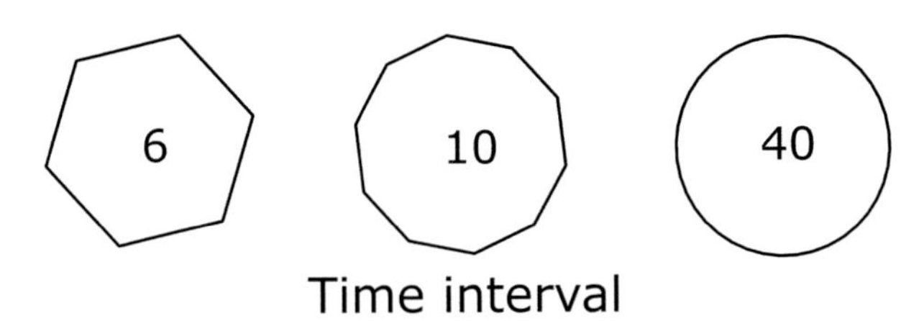

Fig. 5.14 The effect of different time intervals on a calculated circular path

are accelerated to. But when they are accelerated, they will change their position and thus their force effects on each other.

How do we escape this dilemma? The way out is a fine subdivision of time; the changes of positions due to the forces are calculated for short time intervals and then the mass points and their velocities are spatially redistributed according to the forces. Then the calculation starts all over again. The problem of time steps of different lengths for a circular path is shown in Fig. 5.14. The smaller the time steps, the more precisely the circular path is described, with the implication that more and more time steps are needed to complete the circle.

The simplest "brute-force" algorithm is simplified in pseudo code in Algorithm 1.

Start Distribute mass points in space and give each mass point an

initial speed according to the problem.

while *Simulation end is not yet reached* **do**

    for *Every small time interval* **do**

        for *Every i-th mass point* **do**

- Calculate the resulting force on the i-th mass point by summation of all individual forces caused by the other mass points
- Change the speed due to acceleration

        end

- Move all mass points according to the new speeds

    end

end

Algorithm 1 The steps for a simple approximate solution for a multi-body problem

The dynamics of planets, stars or galaxies, which only interact gravitationally, can already be described very well with this recipe. If we know the initial conditions, the locations and velocities of the mass particles, and if we choose a small time step adapted to the problem, we can predict the positions of the particles and their orbits in the future. Of course, there are many mathematical tricks in detail to keep the errors small and to obtain realistic orbits. Dark matter can also be modeled with this algorithm because we only know of one interaction of dark matter with itself and ordinary known matter, namely gravity. If one wants to describe the development of cosmic structures under the influence of dark matter with this model, many mass particles are required in a simulation, and if one wants to do this additionally over cosmic time spans, many small time steps are also required. This results in a problem for the execution time of the algorithm. Because, if we recall the steps of the force summation, then $n(n-1)$ calculations are necessary for n particle. This is an n^2 problem - for large numbers of particles, this is a computationally intensive problem. And this is exactly where the Barnes-Hut algorithm comes in, and it aims to reduce the number of calculations while maintaining satisfactory precision. The basic idea to achieve this is sketched in Fig. 5.15. If we look at particles at a greater distance from a particle for which we just want to calculate the sum of all forces, we first introduce pseudo-particles with a center of gravity coordinate, total mass and center of gravity velocity of the combined particles. After that, we only have to consider the force on this pseudo-particle, saving a lot of arithmetic operations. Of course, the art is in finding the correct center positions, but Barnes and Hut were able to show that this approach works very well with a special space division method.

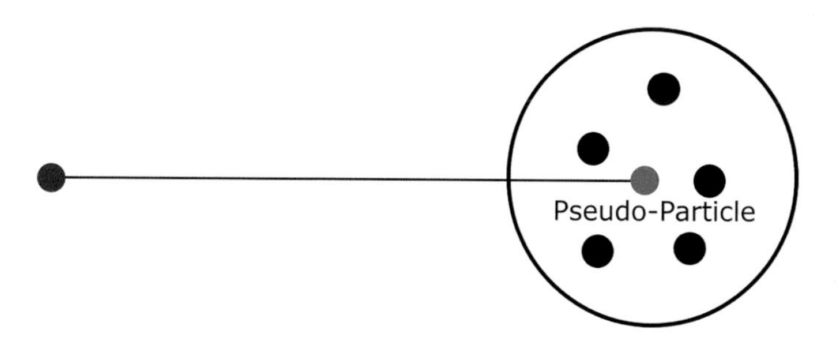

Fig. 5.15 For distant particles, a pseudo-particle is introduced at the center of mass. For the particles in the circle, this particle represents a single, massive particle for the calculation of the force on the red particle

But let us first look at the early days of these simulations because their success paved the way for today's models.

With the advent of ever more powerful electronic computer system in the 1950s and 1960s, astrophysicists were able to simulate the gravitational interaction of up to 16 particles for the first time. At that time, the focus was on the temporal evolution of star clusters; cosmological simulations with thousands or even many billions of bodies were still a long way off. For example, Sebastian von Hoerner (1919–2003), a German astrophysicist, carried out the first direct numerical simulations of globular star clusters on a computer at the Astronomical Calculation Institute (Astronomisches Recheninstitut) in Heidelberg. Only a few years later, attempts were made to investigate the temporal evolution of galaxy clusters using up to 100 particles, the first work of Sverre Johannes Aarseth (born 1934) being worthy of mention. In these works, it was possible to show how galaxy clusters can gravitationally trap field galaxies that were not initially associated with a galaxy cluster. This was achieved on a calculating machine that could perform 100,000 arithmetic operations per second. As a comparison, one has to keep in mind that a current game console can perform about 2,000,000,000,000 arithmetic operations per second, which is a factor of 20 million times more.

With the rapid development of computing power in the following decades, an ever increasing number of particles could be calculated in the simulations, which led to higher resolutions of the structures under investigation and thus to an ever better comparability of the results with the observations. As early as the 1970s, calculations with several 1,000 particles were then carried out to reconstruct the observed large-scale distribution of galaxies. In these early cosmological simulations, due to the low resolutions, that is, small particle numbers, one simulation point represented an entire galaxy. Over the years, the number of mass points in the simulations has steadily increased, so that currently 100 billion simulation particles can be calculated in the largest structure formation simulations. In addition to the physical challenge of modeling, this branch of astrophysics is becoming more and more a top technical achievement in order to be able to answer the questions posed to the model satisfactorily.

But let us first take a closer look at the typical processes of such a purely gravitation-based simulation. The beginning of every n-body simulation is given by the initial model of mass distribution. The first considerations about the properties of a realistic, initial distribution of matter in such a model came from Philip James Edwin Peebles in 1967 (Peebles 1967). The idea of the formation of galaxies from initially small gravitational perturbations of a "matter cloud" goes back to work by the physicist Sir James

Hopwood Jeans in 1902. However, this initial model had great difficulty in explaining the observed structures, as there was no quantitative knowledge of these initial gravitational perturbations. But in 1967, Peebles built the bridge to the cosmic microwave background, which had been discovered two years earlier by Penzias and Wilson. Peebles argued that if the initial perturbations were small enough, in an expanding universe, structures such as galaxies and clusters of galaxies could be formed purely from the gravitational amplification of these perturbations. Of course, Peebles did not then have the information about the fluctuations in the cosmic radiation background that we have today. But the path for a possible formation of galaxies and galaxy clusters, the cosmic web, from a hot universe after the Big Bang was mapped out. It was necessary to determine the exact properties of the distribution of matter at the time of the decoupling of the microwave background, then this initial model of the distribution of matter and its velocities could be further developed purely gravitationally in a computer and the resulting large-scale structures could be compared with the observations. At the end of the 1970s, the time had come. It was possible to carry out the first structure formation simulations, based on initial conditions as observed at the time when the cosmic background radiation was generated, with several thousand particles. Figure 5.16 shows the principle of the initial situation of cosmological simulations. The basic idea is to translate the measured temperature distribution in the microwave background into a density distribution. Let us take a measurement line along the microwave background: The existing small density differences increase over time to stronger and stronger concentrations of matter due to gravity, which is the only attractive force between masses. Metaphorically speaking, gravity acts like a vacuum cleaner, leaving empty spaces free of diffuse matter. The first simulations of structure formation were geared purely to gravitational interactions. The first task was to investigate whether the distribution of the forming clumps in the simulations had a statistical similarity with the observed clumps and their distribution, the galaxies and galaxy clusters. For this purpose, typical distances of galaxies, as found with the Sloan Digital Sky Survey, are statistically compared with those of the simulation.

Figure 5.17 shows results of such a simulation by Sverre Johannes Aarseth, John Richard Gott III and Edwin Turner from 1979. Spherical sections with a radius of about 56 Mpc from the simulation with 4,000 particles for different points in time are shown. The expansion of the universe has been deducted in graphs (A) to (F), so that each section appears to be the same size. Each mass point stands for a galaxy, starting from the initial distribution (A) through the intermediate stages (B) to (E) to the end

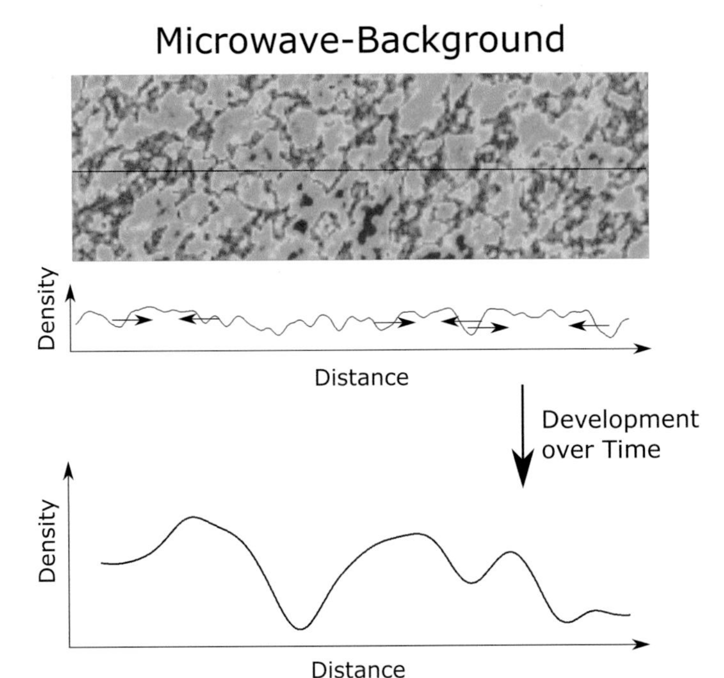

Fig. 5.16 The initial conditions for cosmological simulations. The temperature distribution in the microwave background is converted into a density distribution, here qualitatively represented along a line. Denser regions attract matter from the environment, which leads to an increasing density contrast over time

point (F) of the simulation. With the cosmological parameters chosen at the time, this corresponded to about 17 billion years of development. Over the different time steps, one can see how galaxies gather into clusters and how filament-like structures are formed between the galaxy clusters. The areas without galaxies, the so-called voids, are also clearly visible.

In these early simulations, dark matter was indirectly added to the individual mass points, which were supposed to represent galaxies. This is because each mass point in these simulations is a galaxy with a corresponding amount of dark matter. Without this additional dark mass fraction, structures similar to those observed would not form. Without dark matter, no galaxy clusters could form, as Zwicky already estimated.

Forty years later, the approach is similar, but much more detailed. Not only the resolution but also the physics involved in the simulation has been dramatically increased. Today's simulations calculate with many billions of mass points and take into account the complex physics of gases, star formation and feedback effects from them - even the effects of giant black holes

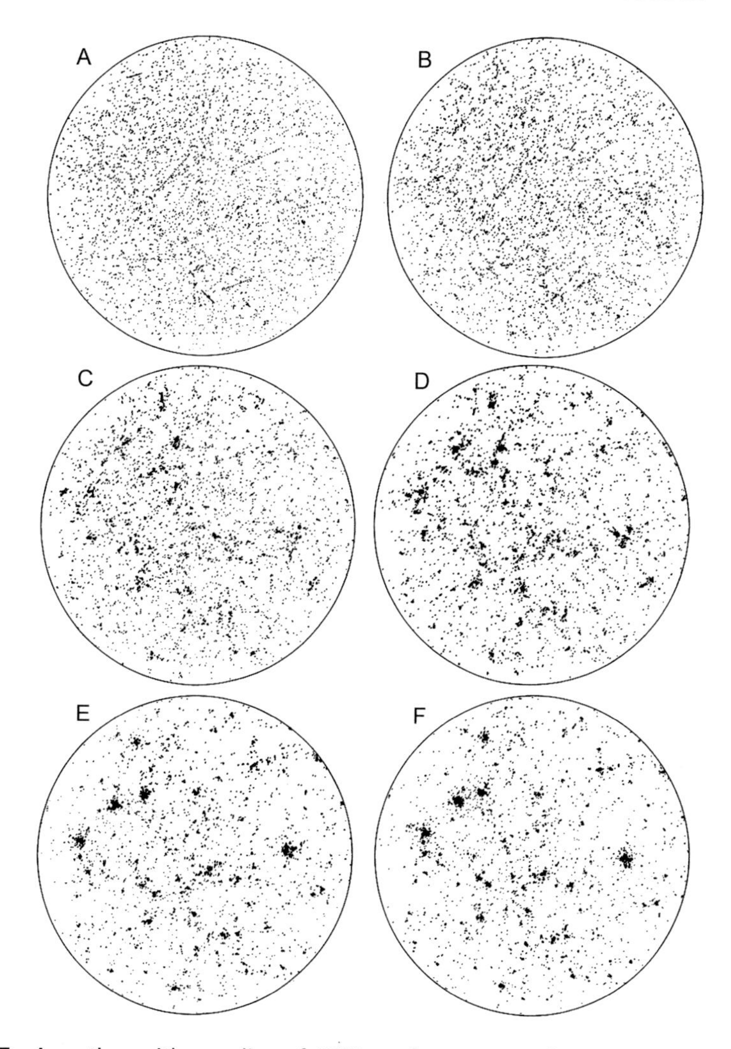

Fig. 5.17 A section with a radius of 56 Mpc of a structure formation simulation for different points in time. The expansion of the universe has been deducted in the graphs (A) to (F), so that each section appears to be the same size. Each point stands for a galaxy. About 17 billion years of development are shown. Source: Aarseth et al. 1979

on their surroundings are part of the calculations. Thanks to hydrodynamics and gravitational physics, we are now able to simulate the formation of the first stars and protogalaxies and show that compact small structures are formed first, which merge into larger and larger ones over time, forming the galaxies and galaxy clusters we have observed. This so-called *bottom-up* scenario, see Fig. 5.18, fits well with the current observation situation.

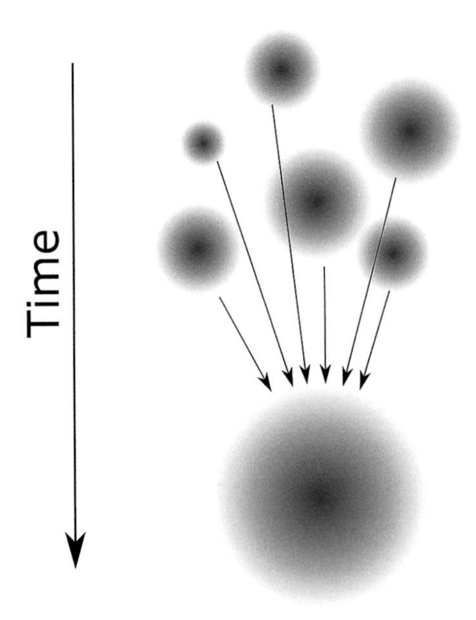

Fig. 5.18 The bottom-up scenario of structure formation in the universe: First, small lumps of matter (of known matter, such as dark matter) formed by gravity - so-called halos. These then merge into larger and larger halos

The mass of these structures, united by gravity, attracts ever-present matter between galaxies and clusters of galaxies and forces it into their mass centers. This primordial gaseous matter, made up of the lightest elements, mixes with the matter within the galaxies and thus forms new germ cells for stars (see Fig. 5.19). In their fusion chains, stars transform this matter into heavier elements, which are then released into the environment again at the end of their lives, partly in the form of hot gases enriched with heavy elements.

Figure 5.20 shows the result of a new, very complex simulation, the Illustris simulation, impressively staged. All in all, the calculated distribution of dark matter, the density of "normal" matter in gaseous form, the temperature and chemical composition of this gas (from left to right) are shown at four distinct points in time (from top to bottom into the past): in the first row to the present time, in the second row about 7.5 billion, in the third about 10 billion and in the last row about 12 billion years ago. The section shown is about 100 Mpc, centered around a galaxy cluster. Obviously, the contrast in density for dark matter, as well as for "normal" matter, increases strongly over the billions of years. This density contrast and the distribution of structures in the simulation results can be compared with the observations. For this purpose, measurements in the entire spectrum of

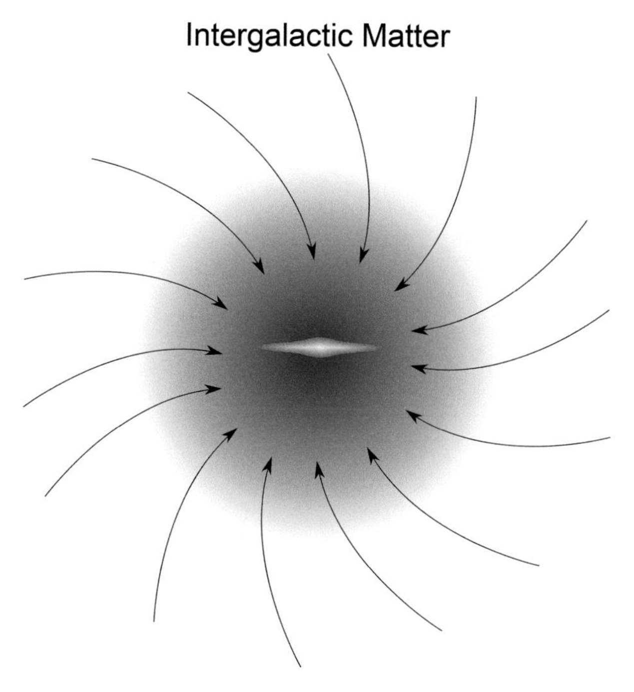

Fig. 5.19 Primordial gaseous matter is gravitationally trapped in halos, mixes with the existing matter and forms new germ cells for stars

light are required. This type of observation, known in the literature as *multiwavelength observations*, is very complex, as it requires the combination of data from different satellites and earthbound observatories. You correctly assume that the concept of dark matter in the simulations has statistically achieved excellent agreement with the observations.

When looking at Fig. 5.20, it is immediately noticeable that dark matter clumps differently than "normal" matter in gaseous form. Since gas is heated by the interaction of the gas particles with each other during compression, also caused by the gravity of dark matter, it does not clump in the same way as dark matter. Dark matter, on the other hand, does not experience any other interaction than gravitational and thus attains a different density distribution. The result is a different density contrast compared to dark matter. Whenever gas in the cosmos contracts due to gravity, stars can form. This behavior is also integrated into modern simulations in a correspondingly parameterized way. In this process, feedback processes occur, which influence the surrounding matter energetically and chemically. On the one hand, decaying stars heat up their surroundings accordingly through feedback processes, for example, supernova explosions. On the other hand, fusion

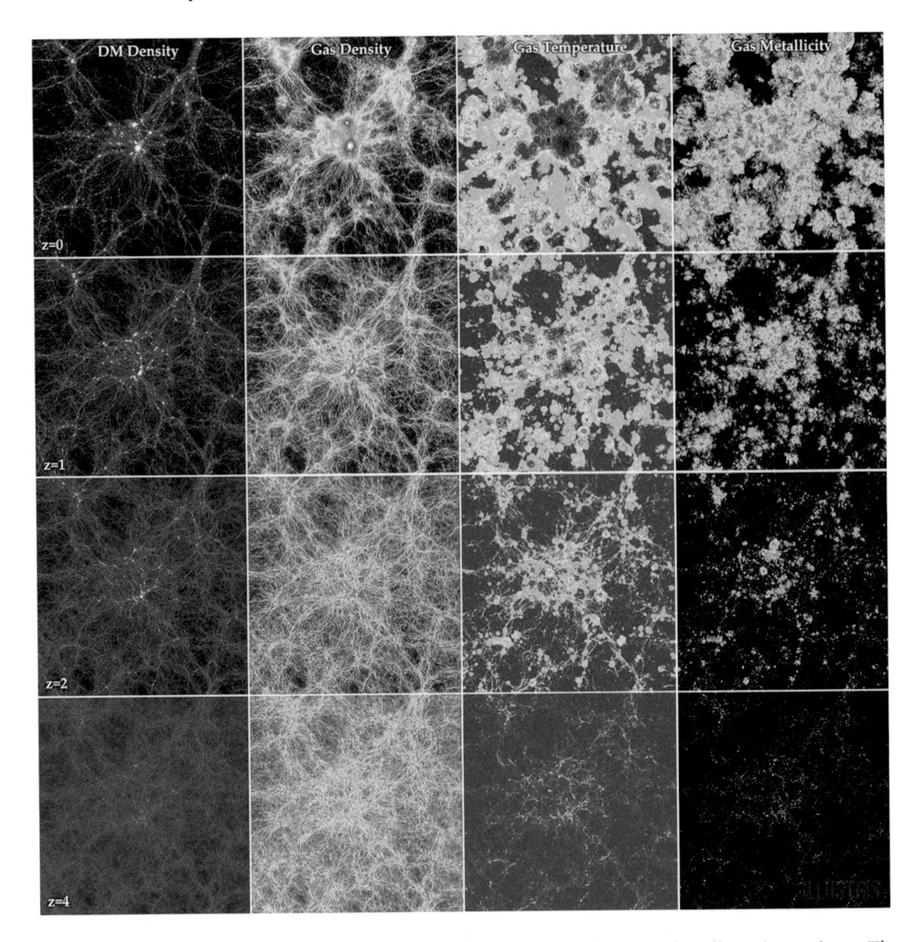

Fig. 5.20 Visualization of a current cosmological simulation - the Illustris project. The distribution of dark matter, gas, gas temperature and the chemical composition of the gas are shown from left to right at four different points in time. The points in time are shown from top to bottom: today, about 7.5 billion years ago, about 10 billion years ago and about 12 billion years ago. The pictures each show an area around 100 Mpc. The expansion of the universe has been deducted so that the structures can be seen better. Source: Illustris Collaboration; https://www.illustris-project.org/media/

reactions in stars produce higher elements, which partly find their way into the surrounding gas at the end of a stellar life. This leads to a change in the chemical composition of matter. In current simulations, this is also taken into account accordingly and the result is compared with the observation. This is also an important test to improve our understanding of the formation of structures in the universe.

How well current simulations can reproduce nature is impressively shown in Fig. 5.21. One can see a comparison of observed galaxies (section of the Hubble eXtreme Deep Field) in the left half and an "artificial" observation of the Illustris simulation in the right half. Even with the naked eye, you can see that the distribution, colors and shapes of the galaxies created in the simulation can statistically describe nature well.

If we remember that there are only about 40 years between the initial cosmological simulations, for example, by Aarseth in the late 1970s, and the results of current cosmological simulations, we can impressively see how extremely successful the concept of dark matter is. One can follow the structure formation on the stage of an expanding universe in a computer, admittedly a powerful one. Without dark matter, there would be no coherent picture of the formation of all the galaxies and even larger structures in the universe.

These simulations are not only very interesting from an astrophysical point of view, but also the demands on the mainframe systems are enormous, so the Illustris simulation presented here required about 19 million CPU hours. On a high performance computer system with about 8,000 computing core, this is a more than 3-month long calculation, which required an enormous amount of RAM of about 25 TB. The model contained about 12 billion particles representing dark and "normal" matter. If we compare this with the simulations of the late 1970s with their 4,000 particles, it can be summarized

Fig. 5.21 Comparison of observed galaxies (part of the Hubble eXtreme Deep Field) left half and an "artificial" observation of the Illustris simulation. Source: Illustris Collaboration; https://www.illustris-project.org/media/

in a nutshell that it was possible to improve the resolution of such simulations by a factor of 3 million. A factor that, overall, also very well reflects the increase in calculations per second per 1,000 dollars on microprocessors over the last 40 years (Source: BCA Research, 2013).

By means of such cosmological simulations, one is also able to calculate hypothetical dark matter signals. In some scenarios, dark matter particles interact with themselves, producing characteristic radiation in the very high-energy gamma-ray range of the electromagnetic spectrum. By observing this characteristic gamma radiation using suitable detectors, it would thus be possible to indirectly detect dark matter, an approach that is currently being pursued and will be described in more detail in this book. Figure 5.22 shows such a hypothetical gamma-ray map of a simulated galaxy cluster, which in this example has about 17,000 dark matter structures (halos). Observations by appropriate observatories can then compare these theoretical maps with real measurements and thus test our still vague models of the nature of dark matter.

Fig. 5.22 Hypothetical gamma-ray map of the largest galaxy cluster in the Illustris simulation. The signal could result from an interaction of dark matter with itself, leaving a characteristic gamma-ray signature. Source: Illustris Collaboration; https://www.illustris-project.org/media/

Problems in the Simulations

Besides all the successes, there are also problems in these simulations. These include the occurrence of too many small structures (halos) in the sphere of influence of larger galaxies. One observable example is dwarf galaxies in the direct vicinity of the Milky Way. The small structures present in the simulation represent a phenomenon that is not observed in nature in this way. Due to the purely gravitational interaction of dark matter with itself and "normal" matter, dark matter tends to form lumps with high central densities. Normal matter, which primarily occurs in the cosmos in gaseous form, is heated when its density increases, which in turn leads to the expansion of these gases. Figure 5.22 gives an impression of how many so-called "subhalos" are formed in halo-like structures such as galaxies and galaxy clusters. In all these halos and subhalos, there are galaxies of different sizes and masses. And especially in the case of small galaxies, dwarf galaxies, such as the Magellanic Clouds in the immediate vicinity of our Milky Way, the simulations statistically achieve an unobservable abundance. More than *ten times* as many dwarf galaxies compared to the observations are created in the simulations.

To solve this problem, the simulations try to integrate models of star formation and the related feedback mechanisms. The main focus is on developing and physically plausibilizing mechanisms that prevent too many stars from forming in the small subhalos between the large dark matter halos. Although the halos are present in the simulations, the feedback mechanisms prevent too much gas from flowing into these smaller halos, thus preventing the formation of new stars. Supernova explosions play a key role in this because they prevent, for example, further gas from flowing into these smaller gravitational pots. Supernova explosions literally "blow" these small halos gas-free. The formation of further stars is thus prevented. An effect that can also be observed well in some galaxies, see Fig. 5.23. In this multiwavelength composite, an optical observation with the Hubble Space Telescope, an X-ray observation with the Chandra Space X-ray Telescope and an infrared observation with the Spitzer Space Telescope are shown together. It shows hot X-ray emitting gas in blue, which was strongly accelerated and heated by an increased supernova explosion rate in the central area of the galaxy. A considerable part of the gas and dust mixture present in the galaxy is ejected into the intergalactic space between the galaxies (shown in red), thus suppressing further star formation in the galaxy.

Fig. 5.23 A composite observation of the NASA Spitzer, Hubble and Chandra space telescopes of the galaxy M82. In blue, you can see hot, X-ray emitting gas, which together with a colder gas–dust mixture (shown in red) is pushed out of the galaxy due to numerous supernova explosions in the center of the galaxy. Source: NASA/JPL-Caltech/STScI/CXC/UofA/

There are successes in the application of this method, although much research is still needed here because every parameter must of course have a physical cause, otherwise simulations would be pure, albeit technically complex "painting." In addition, this process could prevent these dark matter haloes from becoming too dense at their centers, a problem that would otherwise lead to unrealistic dynamics in the galaxies.

Maybe one day, we will be able to observe these halos of pure dark matter. One approach offers an extended description of the interaction of dark matter with itself. This interaction releases radiation in the form of very high-energy gamma rays, radiation that is currently being sought. Figure 5.22 shows such an observation from the simulation.

The simulations of structure formation in an accelerated expanding universe currently achieve a very good agreement with the observed structures in the universe. This is a current, constantly changing field of research, which, with the help of dark matter - in addition to all the problems - is able to provide a fairly consistent picture of structure formation in the cosmos.

5.5 Is It Possible Without Dark Matter?

Up to this point, the only question we have asked ourselves is the amount of dark matter in all the astrophysical objects to support the very successful theory of dynamics. Whether the nature of dark matter is that of a particle, or whether it is the observable effects of other causes, we do not yet know. We have seen that from the small to the large scales, from planets, stars to galaxies and clusters of galaxies, more and more of this dark matter is needed. The "dark" component prevents a failure of Newtonian dynamics or general relativity on large scales. Without dark matter, the observations in our immediate vicinity and the laws derived from them cannot be applied to galactic or even extragalactic scales. This circumstance, however, virtually imposes a different approach: Why not modify the laws of gravity?

This was the approach taken by Mordehai Milgrom, an Israeli astrophysicist, when describing the rotation curves of galaxies in the early 1980s. In his theory, known as MOND (Modified Newtonian Dynamics), he modified Newton's dynamics in such a way that at very low accelerations, such as those occurring in the outer regions of galaxies, an additional term begins to dominate the equations, which can well describe the dynamics in this region without dark matter. This change would occur at accelerations that are 10 billion times smaller than those in the Earth's gravity field. With this change, one would now be able to reconcile the observations on large scales with Newton/Einstein. This change in dynamics is not motivated by an additional observation or experiment. You can see the dilemma in this: Although this modification makes it possible to describe the dynamics on galactic scales very well, unfortunately the cause of this additional acceleration remains in the dark. In fact, the rotation curves of the galaxies can be modeled completely without dark matter, but there is no cause for this modification, so what motivates this modification? What causes this new term in the equations? What "undiscovered" physics justifies this modification? This is the problem with all modifications of the theory of gravity. In any case, MOND has been followed up until today and extended to general relativity. The strength of this theory lies on the galactic scale. In describing rotation curves of galaxies, it can describe observations very well. For large structures like clusters of galaxies or the cosmic web, MOND seems to be unable to describe the observations sufficiently. MOND also needs dark matter for this, albeit on a lower level.

An example of this is the study of the Bullet cluster of galaxies. Figure 5.24 shows a composite image of this cluster of galaxies, a so-called

Fig. 5.24 Bullet cluster of galaxies: Optical range - observed with the Hubble Space Telescope and Magellanic Telescope. Pink - a Chandra Space Telescope observation of the intracluster medium. Blue - a mass distribution derived from the weak gravitational lensing effect of all components in the galaxy cluster. Source: By NASA/CXC/M. Weiss (Chandra X-Ray Observatory: 1E 0657–56)

multiwavelength observation; in the optical range, Hubble Space Telescope and Magellan Telescope, and in pink, a Chandra space telescope observation of the hot X-ray emitting intracluster medium between the galaxies. A mass distribution of all components in the galaxy cluster calculated from the gravitational lensing effect is shown in blue.

Actually, the Bullet cluster represents two clusters of galaxies that are currently in the process of merging. There are several phases in the collision of two galaxy clusters: After a first collision, a series of passages follow in which the two structures are repeatedly spatially separated until they finally form a galaxy cluster. During this process, which lasts several 10 to 100 millions of years, depending on its size and mass, the gas within the galaxy cluster is heated up considerably. This heating is caused by shock waves rushing through the gas, and by the increase in central density concentrations, which lead to increased gas attraction and thus to higher pressures and resulting temperatures. Even galaxies in the center of these clusters eventually merge and form the largest observable galaxies in the cosmos, so-called central dominant (cD) galaxies. In this process, all the galaxies involved mainly feel only the gravitational attraction. The hot gas between the galaxies, shown in pink, with its very low but not negligible density, feels the gas pressure. As a result, it is slowed down and spatially separated from the low potentials of the merging galaxy clusters. These processes create huge shock fronts in the X-ray emitting intracluster medium. These cause shapes similar to those observed in the air around fired projectiles; hence, the somewhat peculiar name Bullet Cluster.

What is special about the investigation of the bullet cluster is the fact that the mass distribution derived from the gravitational lensing effect and that of the hot X-ray gas can be observed so far apart. Dark matter and gas were separated from each other here.

But how can we extract the mass distribution of dark matter in the observation. The effect used is called the weak gravitational lensing effect. The idea behind it is the assumption that galaxies on large scales are statistically purely randomly aligned in space. If galaxies are observed through a large mass accumulation, such as clusters of galaxies, the orientations of the background galaxies appear slightly altered optically due to the large mass of the galaxy cluster and the resulting curvature of spacetime. This results in a statistically significant observable change in the orientation of the galaxies, which makes it possible to draw conclusions about the mass distribution within the galaxy cluster, which acts as a gravitational lens.

Without dark matter, we would expect that the center of mass determined from the weak gravitational lensing effect would coincide with that of the intracluster gas. But the result is quite different. The centers of mass derived from the weak gravitational lensing effect are located far away from the intracluster gas, they are spatially strongly separated from each other. We would expect the same from dark matter because it does not feel pressure like gases, but only gravitational interaction. Gas, on the other hand, is slowed down by the pressure of gas from the other galaxy cluster. This results in the spatial separation of the two components.

MOND fails exactly in the description of such observations. The Bullet Cluster is therefore regarded by many astrophysicists as a prime example for the refutation of MOND. However, it should be noted that the weak gravitational lensing effect carries a very strong assumption: that the orientation of galaxies is statistically random. Certainly, there is still a lot of research to be done to refute this assumption beyond any doubt.

In any case, Mordehai Milgrom pointed out that his modified theory of dynamics requires much less dark matter to explain the observation of the bullet cluster of galaxies.[2] Furthermore, there is a movement within astrophysics that deals with modified dynamics. Thus, a relativistic formulation of the concept was published by Jacob Bekkenstein in 2004. A modification of the dynamics per se is not a forbidden approach - quite the contrary. Newton's theory has already been replaced by Einstein's general theory of relativity in the range of very large accelerations. But the concept here was the curvature of spacetime, which was also confirmed by the gravitational lensing effect. MOND in its original idea reminds too much of an approximation method and the fit parameters contained therein. By skillful choice of parameters, rotation curves can be described, which counterpart these parameters have in reality is not (yet) motivated. But one thing is important to emphasize: There is a need for directions in off-the-beaten-path in research, so that a perpetual critical questioning provides new insights and better approaches.

Summary

After the concept of dark matter brought observations on large scales into line with the theories of dynamics, dark matter also reached the methods of cosmological simulations. This enabled the formation of large-scale structures to be traced. After the initial successes in describing the formation of cosmic networks and the large empty areas in between, the so-called voids, the simulations became increasingly complex and multilayered over time. Today, it is possible to describe the structure formation simultaneously on large cosmological and relatively small galactic scales. Problems in the simulations, such as too many dwarf galaxies, are attempted to be solved with more realistic models of star formation and their feedback processes. Even huge black holes, in the form of active galactic nuclei, influence their surroundings in the current simulations. These are promising approaches, but they always require critical confrontation with the applied models and parameterizations. And it remains a relevant field of research that will provide many new insights.

But there were also new approaches on the theoretical side. Since the 1980s, attempts have been made to modify the theory of dynamics. The aim was to describe the rotation of galaxies without the concept of dark matter. By changing the dynamics at very low accelerations, excellent results in the description of the rotation curves of the galaxies were achieved.

[2]https://www.astro.umd.edu/~ssm/mond/moti_bullet.html

Unfortunately, however, this approach, known in the literature as MOND, does not seem to have the desired effect on large structures such as galaxy clusters or the cosmic web. Also here, not visible matter is needed, even though much less. In addition, the motivation for adapting the dynamics is not based on a cause and thus seems to be hit arbitrarily. Future research will certainly show here whether an adaptation of the laws of dynamics is possible or whether dark matter is found to a sufficient degree in the universe.

In any case, one thing became more and more obvious during these many years of research: The time was ripe for a targeted search for this "dark" component of the universe.

References

Aarseth SJ, Gott III JR, Turner EL (1979) N-body simulations of galaxy clustering. I – Initial conditions and galaxy collapse times. ApJ 228(Part 1):664–683

Barnes J, Hut P (1986) A hierarchical O(N log N) force-calculation algorithm. Nat 324:446–449

Bennett CL et al (2013) Nine-year Wilkinson Microwave Anisotropy Probe (WMAP) observations: final maps and results. ApJS 208:1–54

Dicke RH, Peebles PJE, Roll PG, Wilkinson DT (1965) Cosmic Black-Body Radiation. ApJ 142:414–419

Peebles PJE (1967) The gravitational instability of the universe. ApJ 147:859

Planck Collaboration (2016) Planck 2015 results. XIII. Cosmological parameters. A&A 594(id.A13): 63

6

The Beginning of a Great Search

As more and more observations and their interpretations demanded a dark component of matter, the great search for the nature of dark matter began at the end of the 1990s.

But how do you find something *dark* if our main source of information from space is electromagnetic radiation, that is, light?

Initially, the search was focused on the investigation of known astrophysical objects. It was concentrated on objects that are very faint or completely dark and could perhaps be robbed of their darkness with increasingly better observation technology and indirect observation approaches. Was it possible that dark matter consisted of black holes, brown or white dwarfs? Is it build up by compact, faint objects that are located in large-scale haloes around galaxies, very difficult to observe directly because of their low or nonexistent radiation? If present in sufficient numbers, these objects could reconcile the observed rotation curves of galaxies with the theory of dynamics - a certainly worthwhile advance in the study of the nature of dark matter.

But before we approach the methods and results of the first systematic research projects, let us first look at the properties that a possible candidate for dark matter should have. A profile of dark matter all canditates must meet.

W. Kapferer, *The Mystery of Dark Matter*, Astronomers' Universe,
https://doi.org/10.1007/978-3-662-62202-5_6

6.1 Physical Profile of Dark Matter

We are looking for a candidate for dark matter. The candidate or candidates must:

- have mass,
- *almost* not interact with electromagnetic radiation,
- not have too high relative speeds compared to known matter,
- occur in the space between the galaxies and
- be stable over cosmologically long periods of time.

And what is even more aggravating: It or they must occur sufficiently everywhere in the universe. The search is comparable to the search of a misplaced key. First you look in the familiar places, where you always find it. If the search is not successful, then we move further and further away from the familiar places until we start systematically searching everywhere we have been in the last time. The further we move away, the greater the feeling gets that it makes less and less sense to search there. At the same time, however, the you-never-know feeling becomes stronger and stronger. When we finally find the key, possibly in a very unusual place, there is a strong feeling of clarity. You feel that you have guessed where it is all along, but you wanted to check off all other possibilities before, just to be sure.

It is similar in the search for dark matter. First you concentrate on the familiar places. You examine the usual suspects and compare the wanted profile with their characteristics.

Well, the first property on the profile is obvious, of course, because we remember that mass is the trigger of the problem.

The second property is not fulfilled for many astrophysical objects. The list of objects that are explored by astrophysicists consists mainly of light-emitting entities, which is not surprising considering the main tool of astronomers, the telescope. Only recently, a new window into the depths of space, gravitational wave physics - an entirely new channel for receiving information from space - has opened up. In the next section, we will learn that there are nevertheless some very faint objects that could be considered to be the dark matter we are looking for.

Let us come to the third property, the speed of dark matter. If dark matter would be very fast relative to normal matter, there would be problems in modeling the formation of all the structures such as galaxies and galaxy clusters. These would not be able to form in the way they are observed, there would be an effetct of "washing out" of density contrasts. This effect is discussed in more detail in this chapter.

Furthermore, dark matter must also be present in the huge areas between the galaxies because otherwise we would observe a different temperature and density distribution of the hot intracluster medium in galaxy clusters.

The fifth property is essential. If dark matter were not stable, it would decay into other particles. The energy and mass distributions of the new particles would have clearly observable effects on the shape of structures such as galaxies or galaxy clusters. Indeed, the large-scale structural distribution of matter in the universe would also be completely different.

Perhaps there is a nongravitational interaction of dark matter with what we are familiar with, or maybe dark matter interacts with itself. This question is essential because in this way, we could hope for a signal, for a trace that reveals the nature of dark matter to us.

Although the profile filters out most known candidates, there are some remaining suspects. You will learn more about them in the following sections. You will read how scientists tried and still try to prove to them the dark component of the universe.

6.2 The Search for the MACHOs

We start with a class of known astrophysical objects that represent suitable candidates: The *Massive Compact Halo Objects* (*MACHOs*). In 1986, the Polish astrophysicist Bohdan Paczyński (1940–2007) published a groundbreaking idea for the search for compact dark masses in the halo of our Milky Way. The title of the work is "Gravitational Microlensing by the Galactic Halo," published in the *Astrophysical Journal, Part 1 (ISSN 0004-637X), May 1* (1986). The basic idea of a possible detectability of these objects is based on the gravitational lensing effect. Suitable candidates would be black holes, brown or white dwarfs. In the scientific literature, these objects are referred to by the acronym *MACHOs*. The idea of Paczyński was that these could act as gravitational lenses as they would be present in large numbers in the halo of our galaxy. In doing so, they would amplify the light of a background star, which is located exactly in the line of sight of Earth–*gravitational lens*–background star for a certain time. We want to take a closer look at these objects.

Brown dwarfs are astronomical objects with up to 0.07 solar mass that just cannot be classified as stars. They form a class of objects between gas giants and low-mass stars. Unlike stars, this class of objects does not start hydrogen fusion in their interior due to their lower mass, so they are very faint. The fusion chains in these objects are based on deuterium and lithium, and the resulting luminosities are so weak that a direct observation of a population of brown dwarfs in the halo of our Milky Way is technically impossible at present. Currently, a team led by Jasmin Robert (2016) has been able to prove in a study that the number of brown dwarfs is possibly much higher than assumed after the researchers found about 165 such objects in the close vicinity of our Sun (in a radius of about 50 pc).

The next class of MACHOs is white dwarfs. A white dwarf is the very faint remnant of a *dead* star. White dwarfs usually have the size of the Earth, very high surface temperatures of up to 60,000 K and very high mean densities of up to several tons of matter per cubic centimeter. They represent the final stage of stars with an initial mass of less than 1.44 solar masses. After several billion years, this class of stars can no longer convert enough energy in their fusion chains to counteract their own gravity by means of radiation. They go through a complex final stage in which they release some of their mass into their immediate environment. What remains is a hot white dwarf, a faint object in the mass range between 0.4 and 1 solar mass.

The third class of MACHO objects is stellar black holes. They are formed by stars of greater mass (above 3 solar masses) as soon as in their final stage of evolution no correspondingly high internal pressure caused by sufficiently high energy from an energy conversion process can counteract the gravitational attraction. The final stage of such stars is a black hole, an object that no longer has anything to oppose the gravitational pull of the mass and remains hidden forever from direct observation by astronomers.

It has already been assumed that MACHOs, if they were around the galaxies in sufficiently large numbers, could explain the shape of the observed rotation curves. They would be ideal for motivating the rotation curves of galaxies. They would neither measurably absorb nor scatter the light of distant galaxies, as correspondingly large masses of diffuse dust and gas would do in the halo of our galaxy. Unfortunately, however, one essential property of these objects is precisely one major problem for their detection - their nonexistent or very low luminosity. Other, indirect ways had to be found to detect them. And Bohdan Paczyński was the first to use an already known effect to detect these MACHOs. His idea was based on measuring the brightness of millions of stars in the Small and Large Magellanic Cloud and the nearby Andromeda galaxy M31 or the triangular galaxy M33 over a period of several years. If a compact, massive halo object, such as a white or brown dwarf or a stellar black hole, were to cross the line of sight of an observer–background star of a neighboring galaxy, a brief increase in the brightness of the background star would occur due to the gravitational lensing effect. This increase in brightness is only dependent on the mass of the MACHO, the gravitational lens. In his work, Paczyński calculated probabilities for such an event and the expected course of brightness and based these on the mass distribution of the Milky Way, the measured speed of known objects in our galaxy and the assumed typical masses of MACHOs. Figure 6.1 shows a schematic representation of such an event and the expected brightness curve.

Let us assume that a MACHO is moving at about 200 km/s through the halo of our Milky Way - at right angle to the line of sight of Earth–star.

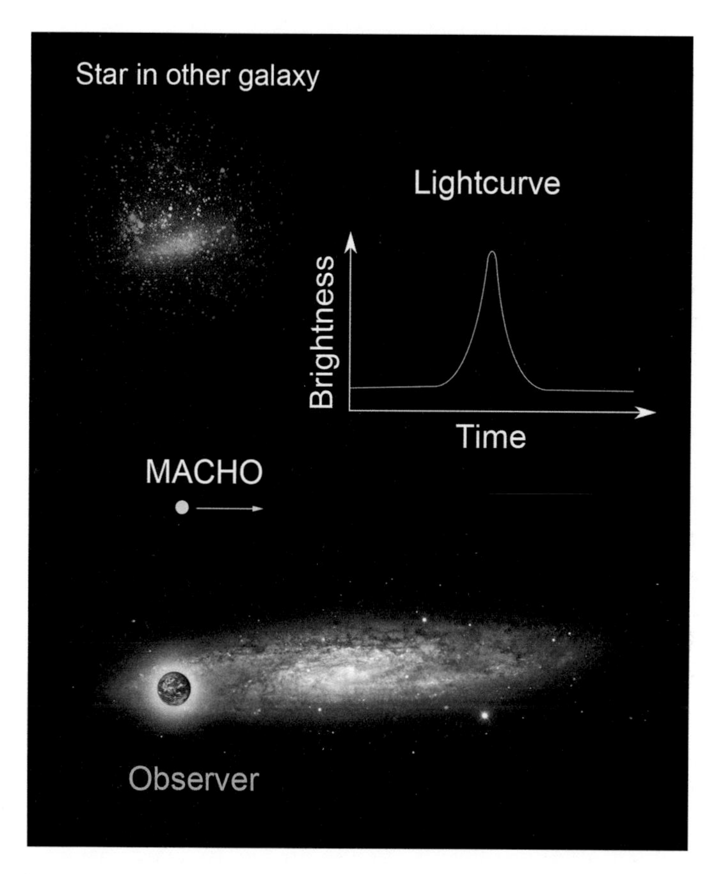

Fig. 6.1 The microgravitational lensing effect caused by a MACHO: If a compact massive halo object (MACHO) such as a white or brown dwarf or a stellar black hole enters the line of sight of an observer–background star of a neighboring galaxy, there would be a symmetrical increase and decrease in brightness, depending on the speed and mass of the MACHO

Furthermore, let us assume that the MACHO has about one solar mass and that the star in the other galaxy is comparable to our Sun in its constancy of luminosity. We would observe a symmetrical increase and decrease of the brightness of the background star over a period of two months. According to Paczyński, we would also be able to observe shorter increase and decrease of brightness as a function of the MACHO masses.

A great idea - but the problem is the number of stars, of which one would have to measure the brightness in short intervals in order to be able to determine the number of these MACHOs purely statistically with adequate accuracy over the frequency of observation events. About 1.8 million stars would have to be studied over a period of one year.

Let us summarize: A program that would observe about 1.8 million stars in the Small or Large Magellanic Cloud (dwarf galaxies in the immediate vicinity of the Milky Way) repeatedly over a period of one year could provide valid statements about the frequency and masses of such MACHOs in the halo of our galaxy.

If the nature of the dark matter in the halo of our galaxy is of the type of massive compact objects, this would be a very elegant way of detection. Assuming that no gravitational lensing events were found, the result would still be a data set of millions upon millions of light curves from stars; a true data treasure for astrophysicists in the field of variable stars, such as the Cepheids, a win–win situation and a very interesting project overall.

Only a few years later, in 1993, the first observation predicted by Paczyński was published by Alcock et al. in *Nature* (issue 365) under the title “Possible gravitational microlensing of a star in the Large Magellanic Cloud.” This observation was made by a collaboration called *The MACHO Collaboration* using a 1.27-m telescope on Mount Stromlo near Canberra in Australia. The aim of the collaboration was to systematically observe about 10 million stars in the Large and Small Magellanic Cloud and in the center of our Milky Way over a period of several years. In 1993, about 1.8 million stars were observed at the center of the Large Magellanic Cloud, about 250 times over a period of 1 year. During the evaluation of the data, a star was discovered that exhibited a light curve that corresponded exactly to a predicted gravitational lensing event. From the amplification of the brightness and the shape of the light curve, one could determine a mass of the gravitational lens of about 0.1 solar mass. In Fig. 6.2, such an observation is shown. Motivated by this discovery, the program was continued until 1999.

There were other such observation programs, such as EROS (Experience de Recherche des Objets Sombres; Dark Object Search Experience) (1993–2002) or OGLE (Optical Gravitational Lensing Experiment), to name but

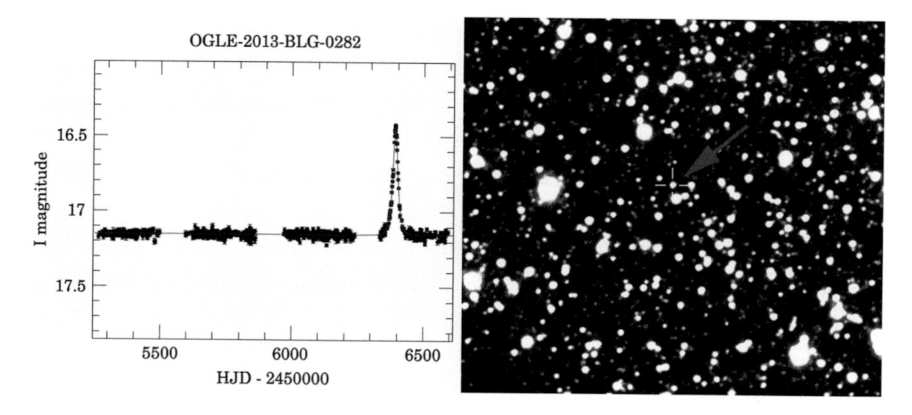

Fig. 6.2 A microgravitational lensing effect event in our Milky Way halo. The observation was made by the OGLE project at Las Campanas Observatory in Chile. Left: The light curve of a star in the Large Magellanic Cloud over a period of 1,200 days. Right: Image of this star. Source: Andrzej Udalski, Michal Szymanski and Szymon Kozlowski, Warsaw University Observatory (Warschauer Universität Observatorium), Warsaw, Poland

two. However, we will return to the results of the search for dark matter using the gravitational lensing effect.

In 2000, the MACHO Collaboration published its final report regarding MACHOs as a candidate for dark matter. After 5.7 years of repeated observations of stars in the Large Magellanic Cloud, 17 gravitational lensing events were measured. However, the duration and properties of the events allowed only one conclusion to be drawn: MACHOs in the mass range of 10^{-7} to just under 0.1 solar mass have a maximum proportion of only 20% of the total mass of dark matter in the halo of our Milky Way. It cannot be ruled out that the MACHOs that caused these 17 events are partly located in the halo of the Large Magellanic Cloud. So at the turn of the millennium, only one conclusion was justified: MACHOs can only account for a fraction of the dark matter. Other collaborations, such as EROS or OGLE, were able to confirm this result over time.

More massive MACHOs in the interesting range from 0.1 to 1 solar mass, such as white dwarfs, can only be completely excluded as candidates for dark matter with longer observation programs, due to the expected lower number of such events. These observations are currently being made and their demands on data evaluation methods are at the limit of what is technically feasible. Thus, after 5.7 years of the MACHO project, some 246 billion individual measurements of stars have been produced; a valuable source for numerous research in the field of variable stars and supernovae. In the course of the years, another branch of astrophysics became aware of this observation method, the extrasolar planet hunters. Today, this method is used, among others, to detect massive extrasolar planets. These become noticeable as a secondary maximum in the gravitational lensing light curves. The gravitational lens is then not a MACHO, but a star with its massive planet, which then leaves its imprint in the light curve of the background star.

It will be interesting to see what future observations in the field of the microlens effect will bring to light. At the moment, there is much evidence that MACHOs are not the nature of dark matter.

6.3 Orphaned Stars in the Depths of Space

In the last two decades, another class of objects in the region between galaxies has been discovered, the so-called intracluster stars. Although they do not quite fit into the general pattern, as they emit light, the mass that is increasingly being discovered in this component cannot be completely neglected. However, their systematic exploration has been technically simply impossible up to our days.

As early as 1951, Fritz Zwicky described in a study of the Coma cluster of galaxies that he could observe a diffuse glow between the galaxies in the cluster (Publications of the Astronomical Society of the Pacific, Vol. 63, No. 371, p.61). Today we know that Zwicky recorded so-called intracluster light from stars between the galaxies on his photographs. Typically, this light corresponds to about 1% of the brightness of the night sky in the optical range and is therefore extremely difficult to study. The light from intracluster stars is widley scattered over clusters of galaxies and becomes brighter towards the central galaxies.

The origin of these stars can be attributed to the formation process of galaxy clusters on the one hand and to the interaction of galaxies with the hot galaxy cluster gas on the other. Interactions of galaxies, which occur during the formation of ever larger structures in a galaxy cluster, transport some stars and interstellar gas from the galaxies into intergalactic space. In the process, these components meet other mass accumulations within the galaxy cluster and are accelerated again. A continuous mixing process until a diffuse distribution across the entire galaxy cluster is observed.

An additional formation process is the stripping of gas from galaxies as they move through the hot intracluster medium. This process is called ram-pressure stripping in the literature. The faster the galaxies move through the hot galaxy cluster gas, the more gas is forced out of them. This is an effect that you experience yourself when you hold your hand out of a fast moving car. The faster the speed, the stronger the pressure of the air on your hand. This effect is also experienced by galaxies moving at 1,000 km/s and more through the thin galaxy cluster gas. New stars can also form in this stripped-off gas, as impressively observed in the blue light outside a galaxy disk in Fig. 6.3.

All these interactions of the galaxies lead to a diffuse star population between the galaxies. Figure 6.4 shows an image of this diffuse radiation in the galaxy cluster Abell 2744 - Pandora's Cluster - taken by the Hubble Space Telescope. The diffuse light of stellar origin is shown in blue. Studies suggest that this stellar component can account for up to 50% of the total

Fig. 6.3 Ram-pressure stripping of the galaxy ESO 137–001. In blue, you can see areas of star formation in the stripped gas of the galaxy. Source: NASA, ESA, and the Hubble Heritage Team (STScI/AURA), M. Sun (University of Alabama, Huntsville)

Fig. 6.4 Intracluster starlight in the galaxy cluster Abell 2744 – Pandora's Cluster. The diffuse light from stars between the galaxies has been colored bluish here. Source: NASA, ESA, M. Montes (IAC), and J. Lotz, M. Mountain, A. Koekemoer, and the HFF team (STScI)

stellar population in a galaxy cluster (Murante et al. 2004). This value depends strongly on where the boundary between the galaxies and the intracluster region is drawn. Nevertheless, in current research, this additional component represents only part of the previously invisible dark matter and is far from being able to account for the masses derived from the dynamics. Especially on the size scale of galaxies, this stellar component is not able to explain the dynamics of the rotation curves. However, it can be used in the range of galaxy groups and clusters to better study the mass distribution of dark matter outside the galaxies. Future observations with better instruments that have better light-gathering capabilities will make it easier to describe the mass distribution in groups and clusters of galaxies using intracluster stars. It is interesting to note here how often one encounters unexpected objects in the supposed "darkness" of the cosmos.

6.4 The Search in the Standard Model of Particle Physics

In addition to the search for macroscopic astronomical objects that may be candidates for dark matter, one field of research in dark matter has increasingly become the focus in recent decades: the microcosm of particle physics. A search began in the particle zoo of the physicists for possible candidates for the mysterious "dark" component of the universe. This is a highly topical field of research that is being pursued in numerous experiments. But before we turn our attention to the experiments and the preliminary results, let us first look at the methods and models of particle physics, so that we can gain an understanding of what kind of objects are actually being searched for.

A Mini Introduction to the Standard Model of Particle Physics

After physicists and chemists of the nineteenth century were increasingly able to describe the properties of matter, it was quantum physics that opened the door to a new era of particle physics around 1900. The idea of quantifying energy made it possible to theoretically predict many previously indescribable observations elegantly and with unimagined precision. Whereas in classical physics, continious energy levels in systems were allowed, new rules were introduced in quantum physics. Like complex game instructions, only certain moves are present in nature. Thus, it was possible to develop a model of the atom that, based on the rules of quantization of radiation, satisfactorily described the behavior of matter in the experiments of those days. Until the 1930s, this model was characterized by its basic components electron, proton and neutron, which formed atoms in the interaction of the elementary forces. A successful model was first formulated by Niels Bohr (1885–1962) in 1913 and basically represents the first quantum physical extension of the preceding atomic models at the beginning of the twentieth century. In Fig. 6.5, the basic statement of the model is sketched. It reads: Electrons in the atomic shell of atoms can only assume discrete energy states, "orbits." These states can be changed by external influences,

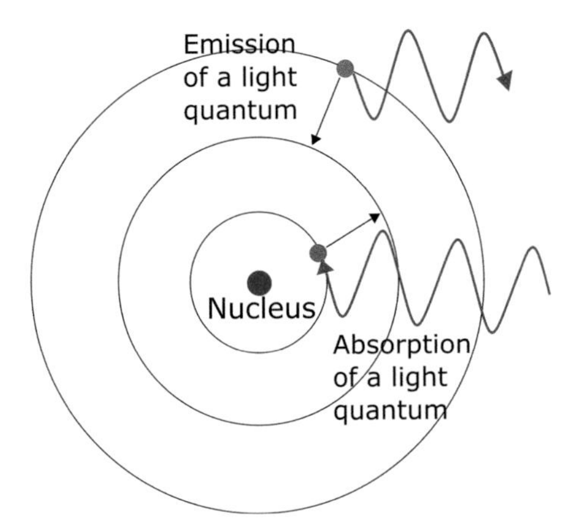

Fig. 6.5 Bohr's atomic model with the postulate that atoms can absorb and re-emit energy in the form of electron excitations caused by light. However, not arbitrarily fine energy differences can be observed, but only discrete ones, which can be described by the rules of quantum mechanics

for example, through excitation by means of a light quantum or through the spontaneous emission of a light quantum. An outstanding achievement of quantum physicists of these early days was the modelling of complex rules of these energy transitions. This was an extremely successful concept, which also led to technological developments such as the laser or semiconductor systems such as the computer.

As successful as these early models were, there were also problems in the description of many observable phenomena and in compatibility with other fundamental physical theories. For example, Bohr's atomic model was not able to describe chemical bonds between the elements, and the observed influence of magnetic fields on the spectra could not be reproduced. Other model properties of quantum physics, such as the spin of elementary particles, were missing to extend the model significantly.

In the following years, additional particles were discovered, such as the neutron in the atomic nucleus theoretically postulated by Sir James Chadwick (1891–1974) in 1932. In the same decade, Carl Anderson (1905–1991) and Seth Neddermeyer (1907–1988) were able to demonstrate a novel particle in the interactions of cosmic rays with matter in the high layers of the atmosphere. This particle, which is similar to the electron in certain properties, the muon, has a mass about 200 times greater than that of the electron and, in contrast, a very short average lifetime in an isolated state of only about 2 μs. It then preferentially decays into an electron and a muon- and electron-neutrino.

Another milestone in the understanding of the structure of matter was achieved with the scattering experiments of electrons on protons.

Scattering experiments are historically the gateway of particle physics to the inner structure of matter. For example, a thin film of an element is bombarded with particles such as electrons or protons. The scattering of the particle beam can be measured with detectors placed around the experiment. Many particles of the beam do not experience any path deflection. Some are deflected, i.e. scattered. From the statistical distribution of the geometry of the measurements around the experiment, one can draw conclusions about the internal structure of the bombarded matter.

First predicted theoretically by André Petermann (1922–2011) and Murray Gell-Mann (1929–2019), further structures within the proton were discovered in the 1960s in experiments in which electrons were fired at atomic nuclei. It was recognized that, in addition to the most elementary particles known at the time, the electrons, muons and neutrinos, there are other elementary units in nature. These building blocks, known as quarks, form so-called hadrons in various combinations, which are held together by a very strong force, the so-called strong interaction. These include neutrons, protons and also mesons.

In the mathematical modelling of these most elementary objects, it was recognized that by choosing the appropriate symmetries, a good model for describing elementary particles could be achieved and that it was possible to make theoretical predictions about particles, which were later discovered in the experiments.

In nature, there are various symmetries that we encounter in all areas. Take a pendulum, which we place on a laboratory table and whose movement we measure. It will make the same movement if we move our laboratory table by one meter and restart the experiment with the same deflection. Starting the pendulum experiment this afternoon or this evening will not change the result of the experiment. There is a symmetry in space and time. Another example is the movement of an electron in a magnetic field. The electron follows a certain path according to the start parameter. A positively charged particle of the same mass with the same start parameters will follow a mirror-symmetric path. Particle physics knows many such symmetries, not only spatial, but also charge symmetries or abstract mathematical ones. The success is based on being able to describe the properties of elementary particles with a set of few symmetries. By means of these symmetries, one can search very specifically for new particles in the particle experiments.

The basic-set of elementary particles with their properties and the possible interactions between them is called the Standard Model of particle physics, which got its name from a lecture at a conference in the mid-1970s by a Greek-born physicist named John Iliopoulos entitled "The Current Model of Particle Physics.

The term particle is certainly misleading on this scale because in our everyday world we associate it with a small ball, a tiny ball with its Newtonian properties, which is not applicable in the world of elementary particles. It is worth remembering the double-slit experiment, which constantly reminds us that matter has wave- and particles-character at the same time, depending on the experimental arrangement chosen. The theory that describes nature on such scales, quantum mechanics, provides us with an instrument that allows us to describe this world and make predictions about it. But it is completely decoupled from our everyday perception. Let us take the hydrogen atom, the most common element in the universe, as an example. If we describe its only electron in the shell, in quantum mechanics, we speak of a probability of the electron being in a location in the shell of the hydrogen atom. In contrast to the atomic model of Niels Bohr, with its almost planetary like electron orbits, a quantum mechanical description of the location of an electron in the orbit of a hydrogen atom is somewhat more complex. The key to understanding quantum mechanics seems to be the realization that quantum mechanics is a statistical theory. A measurement is a kind of lottery draw in the figurative sense. Of all possible, probable and improbable truths, we draw a reality at the time of a physical measurement. Figure 6.6 shows quantum mechanical calculation for an excited hydrogen atom. The electron has locations in the atomic shell with high (in red) and low (in grey) probability density. Depending on the excitation, there are different shell configurations, i.e. probability distributions. The electron, drawn as a ball, corresponds to a hypothetical measurement of the location at a given time. The locations of the electron in different configurations may be very different from Bohr's electron orbits. That just shows impressively how even simple states of matter are highly complex in their describtion in the world of particle physics.

So let us take the term "particle" not too literally at this point. It is only meant to be a placeholder for an elementary piece of nature, which is predictable for us by means of the laws of quantum mechanics. These laws are like nature's game instructions for arranging, changing and transforming matter - although a very complex and far from completely formulated.

In Fig. 6.7, the actors in this model of nature are now summarized. The six different quarks known to us are shown in violet. These quarks form in different combinations, for example, protons, neutrons or mesons. In green are the six different leptons known to us, including the electron and various

Fig. 6.6 The probability of an electron being in a hydrogen atom that is in an excited state. In red, places with a high probability of residence. From blue to grey, one observes lower and lower probabilities of residence. The electron, drawn as a ball, corresponds to a hypothetical measurement of the location at a given time. Source: Dr. Sabine Kreidl

neutrinos. On the other hand, the leptons also include muons and tau leptons, which are very similar to the electron but have a very short life span in free form.

To model the various observed interactions between the particles, so-called exchange particles i.e. bosons are used. A well-known representative is the exchange particle of the electromagnetic field - the photon, which can be found in the red area of Fig. 6.7. To model the strong interaction that holds the quarks together in the hadrons, we find the gluon as an exchange particle, and for the weak interaction, the Z and W boson. This interaction works on small scales. The Higgs boson is the excitation of the associated Higgs field, which, according to the model in the complex game

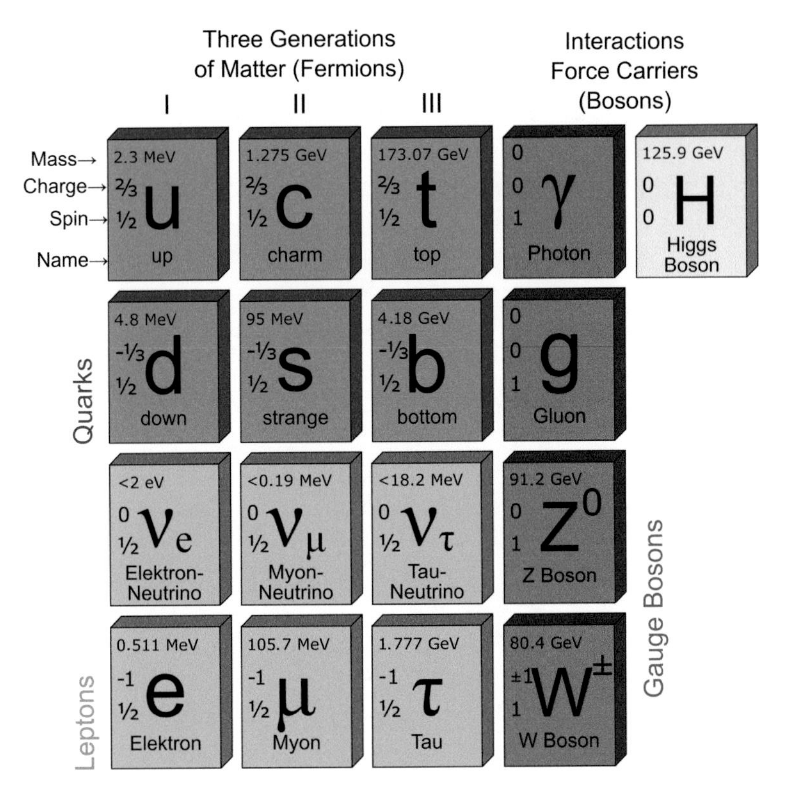

Fig. 6.7 The current Standard Model of particle physics. Source: By MissMJderivative work: Polluks (Standard_Model_of_Elementary_Particles.svg) [CC BY 3.0 (https://creativecommons.org/licenses/by/3.0)], via Wikimedia Commons

instructions, gives all the particles their mass. In this process, particles interact with the omnipresent Higgs field in such a way that we can interpret this interaction as inertia or, indeed, as mass. In particle physics, mass is not expressed in kg, but in electron volts (eV), a unit of energy. This goes back to Einstein's famous formula for the equivalence of rest mass and energy, $E = mc^2$, and is indicated with unit factors Mega (M) or Giga (G).

When one hears of the physicists' intention to develop a "world formula," this actually means the unification of the description of the four interactions. Currently, gravity is described with the general theory of relativity, the strong interaction with the so-called quantum chromodynamics and the electromagnetic and weak interaction with the electroweak interaction. With the exception of gravity, these are quantized field theories. A unification to a quantum field theory of gravitation would be a possible breakthrough towards a "world formula". The goal is to model all interactions with only one theoretical building. Whether this "one" model, this "world formula," can ever be developed remains open. The biggest hurdle is probably a unification of quantum theories with gravity.

But back to the question of dark matter in the Standard Model of particle physics. With this rule-set for particles and their interactions, you will be able to model the known matter in the universe. Yes, there are open questions in the Standard Model, but it is currently the best tested model for describing known matter that we have available here and now and that makes extremely accurate predictions. The crucial question in the search for the "dark" side of the cosmos is: Is there possibly dark matter in the standard rule-set of particle physics? Or to put it another way: is there one known particle in the Standard Model that is *the* candidate?

Once a Hot Candidate - The Neutrino

What if the Standard Model of particle physics already contained a good candidate for dark matter? What if it had already been found at the subatomic level in particle physicists' experiments? Particles that hardly interact directly with the matter we know of and that are very long-lived, but nevertheless have a measurable mass and thus leave their signatures in the dynamics of astronomical objects and are thus candidates for dark matter.

A possible candidate was postulated in 1930 by the Austrian physicist Wolfgang Ernst Pauli (1900–1958). Originally, he called it a neutron and it was not discovered in particle physicists' experiments until more than 20 years after this theoretical work. It has become known as the neutrino. But how did the postulate of this neutrino come about? What prompted Pauli to take this step? And is it a candidate for dark matter? In order to answer these questions satisfactorily, we must look at particle physics in the first half of the twentieth century.

The Power of the Conservation Laws

The history of the neutrino is closely interwoven with the successes of the symmetry-based conservation laws of physics. Initially, conservation laws were known mainly in classical mechanics. With the advent of thermodynamics, the concept of conservation of energy was coined by the German physician Julius Robert von Mayer (1814–1878) for closed systems. The basic idea is that a constant amount of energy in a closed system can be converted into various forms, but is never lost. Influenced by the triumphal procession of steam-driven machines of those days, a successful theory of energy conversion processes was the key to ever more efficient technology. All types of thermodynamics, whether technical, chemical or statistical, are essentially based on the conservation of total energy in the closed systems under consideration. Energy can be converted into various forms (for example, into mechanical or thermal energy), but by no means does energy disappear from a closed system. In the innumerable apparatuses, such as combustion engines, refrigerators or air conditioners, we can observe the success of the conservation of energy every day anew. But with the advent of theoretical research into atomic processes at the beginning of the twentieth century, the law of conservation of energy was confronted with a new type of physics, atomic physics. A new branch in which mechanistic laws no longer seemed to be valid.

At the end of the 19th century, Antoine-Henri Becquerel (1852–1908) discovered that uranium salt was capable of blackening photographic plates. This substance had to emit a kind of radiation. Ernest Rutherford (1871–1937) was able to show a little later that there were two types of radiation, which he named α- and β-radiation. The distinction described the different penetrating capacity of radiation through matter. For example α-radiation can be completely shielded by a sheet of paper, whereas β-radiation requires more matter, such as acrylic glass a few millimeters thick. Rutherford discovered that α-radiation consists of helium nuclei (composed of two protons and two neutrons each) - a particle radiation emitted by certain radioactive decay processes. Depending on the source material, the resulting α-radiation has different energies. Thus, spectroscopic examination of the radiation allows to draw conclusions about the respective source material (radionuclide). On closer examination of the occurring energies, it was possible to show that here the conservation of energy at subatomic level was not violated. The mass, which is converted into kinetic energy by nuclear decay, is divided into α-particles and the original nucleus.

But then the physicists began to deal with the β-radiation, an energetically stronger type of radioactivity, which is produced by another type of nuclide, so-called β-emitters. A distinction is made between two types of β-radiation, β^+ and β^-, due to the different initial properties of the nuclide. The β^--emitter, see Fig. 6.8, is a nucleus with an excess of neutrons. A neutron decays into a proton and emits an electron and a neutrino, more precisely an electron-antineutrino. If the nuclide has an excess of protons, a proton is transformed into a neutron and the radiation emitted consists of a

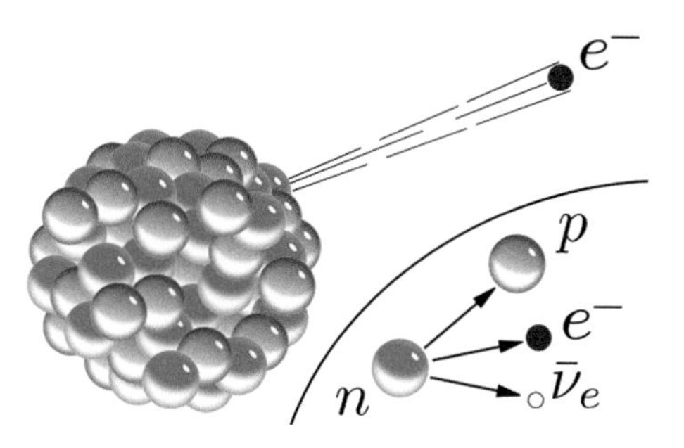

Fig. 6.8 β^--radiation. A neutron decays into a proton and emits an electron and a neutrino (electron-antineutrino). Source: Wikipedia (License: Public Domain)

positron (the antiparticle of the electron) and a neutrino, more precisely an electron-neutrino.

But back to Wolfgang Pauli, who postulated the neutrino. In his time, physicists were only able to observe the electron in β^--radiation, but not the antineutrino. After checking the energy balance of the final system, it was very quickly clear that the law of conservation of energy was violated here. It seemed that energy was being lost.

This led to a deep problem of conservation laws in particle physics. Similar to the phenomenon of dark matter, Wolfgang Pauli in 1930 - assuming the conservation of energy - postulated a hypothetical particle based on the energy distribution of electron radiation, the neutrino. It was named somewhat later by Enrico Fermi (1901–1954), an Italian Nobel Prize winner in physics. This particle, which had not yet been discovered in the 1930s, was eventually to save the energy conservation law of particle physics in β^--decays.

The parallel to dark matter is very interesting here. Although on a completely different physical scale, the analogy of the problem and the chosen solution are striking. The approach was crowned with success because 26 years later, Wolfgang Pauli's postulated neutrino could be discovered. This was achieved by the American physicists Clyde Lorrain Cowan Jr. (1919–1974) and Frederick Reines (1918–1998) with the neutrino experiment named after them. Their approach was to let the neutrino interact with a proton. This reaction could then, according to the theory, produce a neutron and a positron. The positron would then be annihilate with an electron into two high-energy photons, which could be detected. A water tank was chosen as the experimental setup because the proton nucleus of hydrogen in the water molecules proved to be an ideal target. The neutrino source was a nuclear reactor, which, according to theory, constantly emits a very high number of neutrinos. In 1956, the two physicists were then able to observe the high-energy photons in the water tank exactly according to the model and thus indirectly infer the postulated neutrino radiation. The cross section of the neutrinos they measured, i.e. the probability that they would interact with a proton, was also very close to the theoretically predicted value. Frederick Reines received the Nobel Prize in Physics in 1995 for his achievements in the field of neutrino physics.

The law of conservation of energy at the subatomic level was saved, a triumph of theoretical physics.

Neutrino Sources in the Cosmos

In the vastness of the cosmos, the fusion reactions inside stars are strong sources of neutrinos. In the Sun, for example, around 2×10^{39} neutrinos are produced per second, which causes a neutrino flux of about 100 billion neutrinos per square centimeter and second here on Earth. Because neutrinos hardly interact with other particles, they are ideal for studying processes inside our Sun. On their way from the interior of the Sun to us, they carry unaltered information about their place of origin. Large-scale experiments such as the Super-Kamiokande in Japan (see Fig. 6.9) or the IceCube in Antarctica (see Fig. 6.11) are used to detect cosmic neutrinos on Earth and to study their physical properties. Water, whether frozen or liquid, plays the role of the interaction medium. The resulting light flashes during electron-antielectron (positron) pair annihilation are then studied.

But the measurements of the neutrino signals once again led the particle physicists' models into a dilemma. Only about a third of the theoretically predicted neutrino flux could be measured here on Earth. For many years, particle physicists worked on solving the problem of the missing solar neutrinos. One promising approach was the hypothesis that the neutrinos created inside the Sun are transformed into other forms of neutrinos on their way to us. The fact that neutrinos can transform into different types of neutrino was quite possible theoretically, i.e. in the models of particle physicists. However, this transformation only worked if neutrinos had a low mass. And after several decades of neutrino physics, the Sudbury Neutrino Observatory (SNO; Sudbury-Neutrino-Observatoriums), see Fig. 6.10, was able to support these theoretical models with observations in 2001 and 2002. The SNO was the first to measure the predicted different neutrino fluxes here on

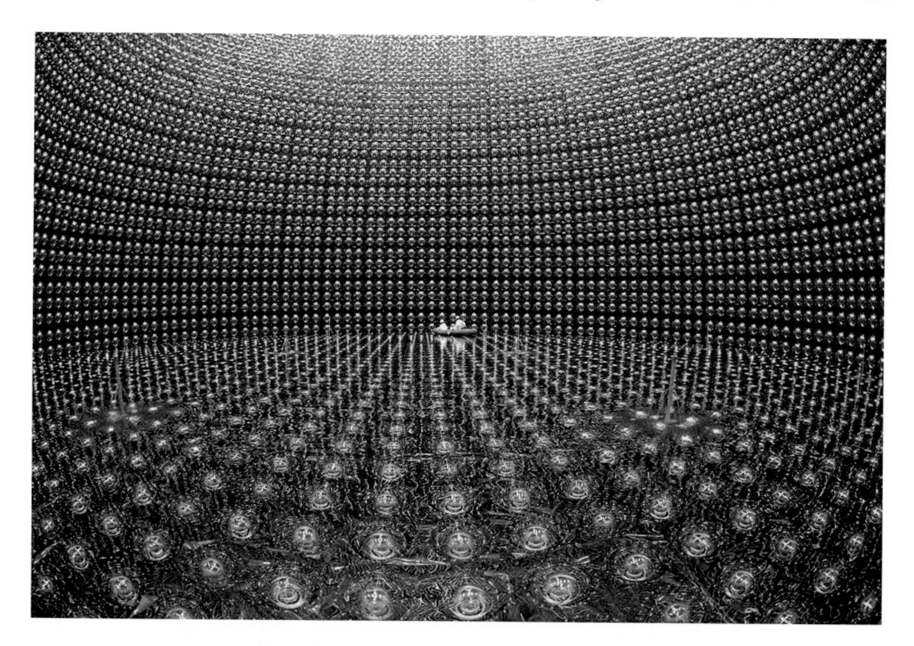

Fig. 6.9 Super-Kamiokande: A neutrino detector in a decommissioned mine in Japan, where light flashes due to the neutrino–proton interaction are measured in a huge water basin. Source: Kamioka Observatory, ICRR (Institute for Cosmic Ray Research) - The University of Tokyo

Earth. About one third of the neutrinos are electron neutrinos. The rest are so-called muon and tau neutrinos.

This had shown that neutrinos transform into different types during their journey from the Sun's core, the site of the fusion reaction, to the surface of the Sun and then to the depths of the neutrino observatories on Earth. This process is known in literature as neutrino oscillation. In addition, this observation allowed us to assign an upper mass limit to the neutrino. But most important in the context of the search for dark matter is that it was possible to show that neutrinos have mass and that they therefore have to be taken into account in the mass budget of the universe.

At present, the properties of neutrinos and their sources are being studied in detail in various experiments around the world. A supernova explosion (1987) has been identified as a neutrino source. Using giant neutrino detectors such as the IceCube in Antarctica (see Fig. 6.11), neutrinos from deep space are being studied and it is hoped that this will help us to find out more about their formation processes. Neutrino physics is a highly topical branch of research.

Fig. 6.10 The Sudbury Neutrino Observatory near Sudbury, Canada: A tank at a depth of 2,000 m below the Earth's surface measures the traces of neutrinos. It is filled with 1,000 t of so-called heavy water, in which the heavier hydrogen isotope deuterium is highly enriched. Source: Photo courtesy of SNO

Neutrinos as Dark Matter

The question of the mass of neutrinos has not yet been finally clarified. Previous experiments have at least been able to establish upper limits. For the electron neutrino, for example, an upper mass limit of about 3.56×10^{-36} kg. This is a mass that is about 450,000,000 times less than that of the proton. Current research in this field, such as the KATRIN experiment (KArlsruhe TRItium Neutrino), may measure a more accurate mass for neutrinos in the next few years.

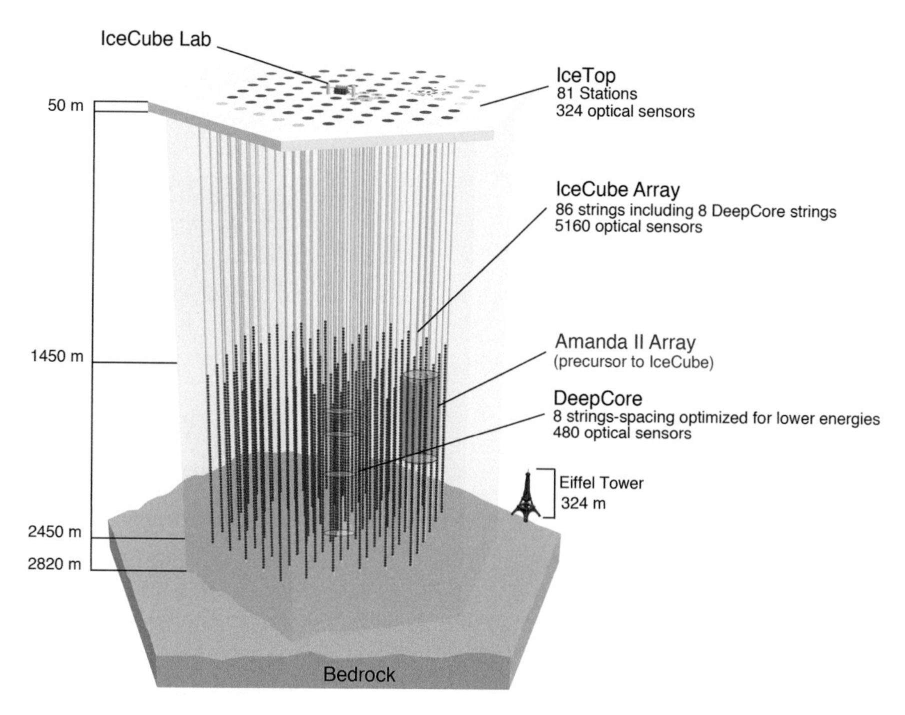

Fig. 6.11 The IceCube neutrino detector in an ice shelf in Antarctica. Light flashes are observed due to the neutrino–proton interaction. As a size comparison, the Eiffel Tower is shown in the lower right-hand corner. Source: IceCube/University of Wisconsin-Madison/NSF

The extent to which neutrinos are candidates for dark matter can be better understood if we consider the possible influence of neutrinos on the formation of the large-scale structures. Dark matter structure formation simulations that take into account the properties of neutrinos are not able to explain the observed large-scale structures. Due to the high velocities of the neutrinos, they do not generate enough high density locations in these simulations, which are necessary for the adequate formation of galaxies and stars. This effect is also known as *gravitational washing* and means that due to the high speeds, no such strong gravitational potentials can be formed. It is similar to a gas mixture. The lower the average velocity of the gas particles, the lower the temperature of the gas, which causes the gas density to increase. This is also the case with matter in space, the denser the matter is concentrated in space, the more it "vacuums" the environment empty, resulting in a high density contrast. But if the average speeds are higher, these density

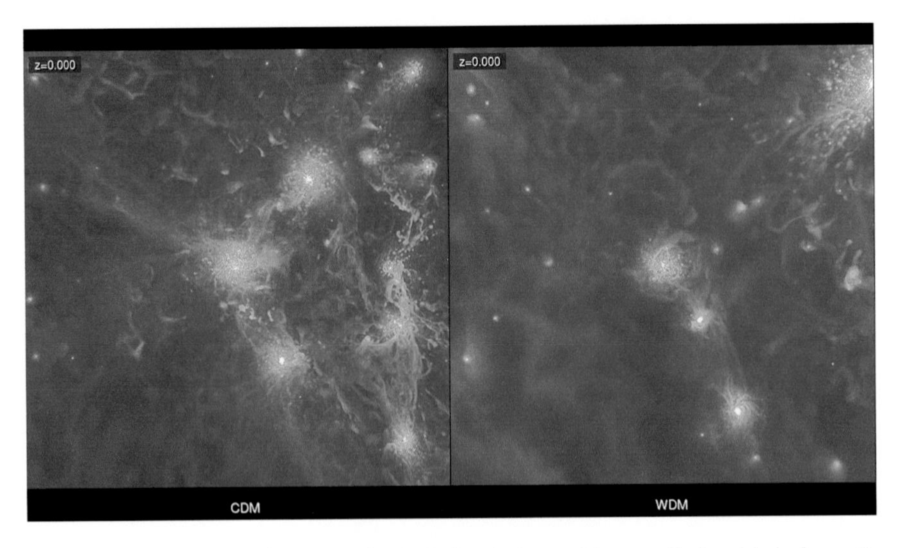

Fig. 6.12 Comparison of a cosmological simulation with so-called cold dark matter (**a**) and warm dark matter (**b**). The difference lies mainly in the initial distribution of the velocity of the dark matter simulation particles. Source: Image courtesy of the CLUES collaboration, https://www.clues-project.org

contrasts "wash out." In Fig. 6.12, on the left, the distribution of gas in a "successful" structure formation simulation is shown, and on the right, there is a simulation with dark matter, partly consisting of particles with high initial velocity (so-called *warm dark matter*), a mixture of dark matter with a velocity of standard matter and a large, neutrino like fraction.

The germ cells of stars and galaxies do not form in sufficient number. We can see how the neutrino particles blur structures and produce a result that contradicts our observations of the cosmos. From the perspective of current cosmology, neutrinos can only contribute to a very small part (a few percent) of the mass budget of the universe. In other words, dark matter consists of neutrinos to a very small proportion, less than 2%. Therefore, from today's perspective, the search must continue.

6.5 Dark Matter Beyond the Standard Model

Looking at the other elementary particles in the Standard Model, none, except the neutrino already considered, represents a possible candidate.

The quarks build up hadrons, which are clearly visible in the electromagnetic spectrum due to interactions and therefore cannot be considered as an explanation for dark matter. The free neutron decays in about 15 min into a proton, electron and electron-neutrino, unless it is much more likely to be absorbed by an atomic nucleus before. The decay components in turn emit observable light when interacting. Free quarks have not yet been observed and whether they exist is not yet theoretically clear. Similarly, leptons, except for neutrinos, would reveal oneself by observable radiation. This class of elementary particles, the electron, muon and tauon, has an electrical charge that emits radiation when accelerated in magnetic fields and is therefore visible to our instruments. Not to mention the stability of elementary particles, free muons and tauons are not stable and decay into stable neutrinos and electrons. So there is no place in the Standard Model of particle physics. One has to look outside the Standard Model and needs an extension of the Standard Model.

The idea of Supersymmetry and the Weakly Interacting Massive Particle (WIMP)

A possible extension of the Standard Model is supersymmetry. This is based on a hypothesis of particle physics, which assigns a partner with the same properties to each elementary particle, with the exception of $\frac{1}{2}$ shifted spins. Each elementary particle would have a corresponding supersymmetric partner particle. Based on their spin, the elementary particles can be divided into bosons with integer and fermions with half-integer spin. But before we look at the significance of hypothetical supersymmetric partners of bosons and fermions for the dark matter model, we should briefly consider the significance of spin in particle physics.

The Spin of Elementary Particles

Let us look again at Bohr's atomic model, see Fig. 6.5. This model states that electrons can absorb or emit light when changing from one discrete energy level to another. In this model, these special transitions are the cause of the observable lines in the spectra of objects, whether in the earthly laboratories or in the light of the stars.

However, if, for example, light-emitting gases are observed in the laboratory under the influence of an externally applied magnetic field, a splitting of lines can be observed. One observes several lines with a magnetic field, where without magnetic field only one could be detected. This effect is known in the literature as the Zeeman effect, named after Pieter Zeeman (1865–1943), a Dutch physicist. He reported in 1896 in a publication of the Physical Society of Berlin about the splitting of a so-called D-line in the spectrum of a piece of asbestos soaked with common salt in a gas-oxygen flame as soon as it was put between the poles of an electromagnet. He himself spoke of a magnetisation of the spectral lines. Zeeman used a theory by the Dutch mathematician and physicist Hendrik Antoon Lorentz (1853–1928) to explain what was observed. Lorentz motivated the cause of light as the oscillation of small electrically charged mass particles.

The introduction of a theoretical characteristic of the electron, the so-called electron spin, by the physicists Samuel Goudsmit (1902–1978) and George Uhlenbeck (1900–1988) in 1925, however, made it possible to explain this observed line splitting elegantly and to causally assign the effect to the electrons in the atoms. Their achievement consisted in recognizing that the theoretical characteristics of the spin gave an additional degree of freedom in the description, which made possible an exact prediction of the energy distribution in the spectrum. The electron was ascribed a property in the manner of a classical intrinsic angular momentum. For a better understanding, one could say that the electron rotates around its axis, creating a magnetic field. This magnetic field in turn interacts with the externally applied magnetic field, which leads to a fanning out of the energy levels of the lines. The fact that discrete bylines are created is explained by the fact that nature only "allows" energies to be quantized at microscopic level, i.e. only in discrete jumps, and hence the name quantum physics. However, the electron does not really rotate. This analogy is due to our classical view of magnetic fields and has only limited validity in the subatomic range.

In any case, the model was so successful that many characteristics of quantum mechanical systems could be excellently explained by this additional property called spin. Elementary particles possess this model parameter spin in one-half and one-integer values, see Fig. 6.7. Quarks and leptons are attributed a half-integer spin, bosons, which are responsible for mediating the interactions between the particles, an integer spin. These two classes can each be described with an adequate theory, the Fermi–Dirac statistics or the Bose–Einstein statistics, which statistically model observable properties of objects consisting of many such particles in a statistically manner. Thus, spin is the third property, besides mass and charge, that characterizes matter in the model of particle physics at the elementary level.

What is a WIMP and Why is it an Ideal Candidate for Dark Matter?

Supersymmetry postulates hypothetical elementary particles and thus assigns each boson or fermion a supersymmetric partner. The basic symmetry of this model is based on a special group, the so-called Poincaré group, named after the French mathematician and physicist Henri Poincaré (1854–1912). Each spin-$\frac{1}{2}$ particle is assigned a supersymmetric spin-0 partner particle and each spin-1 particle is assigned a supersymmetric spin-$\frac{1}{2}$ partner. This doubles the amount of elementary particles, and even more elementary particles are added for consistent supersymmetry. The supersymmetric partner of leptons and quarks is distinguished by physicists by putting an S before its name, i.e. the electron becomes a selectron or the quark becomes a squark. The bosons get a postfix, namely *–ino*, for example, the photino. Some of these supersymmetric particles would be ideal candidates for dark matter.

The section title contains the question "What is a WIMP?" Now the abbreviation stands for *Weakly Interacting Massive Particle* . From the theory of supersymmetry, partner particles such as gravitino or, as combinations of supersymmetric boson partners, so-called neutralinos are possible. According to the theory, such particles would be stable, would have no electric charge, would be massive, would interact via gravity and would also not too fast - virtually ideal candidates for dark matter.

At this point, it must be emphasized that there are other hypotheses in particle physics that predict even more exotic WIMPs. However, supersymmetry currently represents the most promising extension of the Standard Model, even though this is still a largely untested hypothesis.

The Axion

Similar to the phenomenon of dark matter, there is an extremely successful theory in particle physics, but it cannot satisfactorily explain all the observations of the systems it is supposed to model: quantum chromodynamics. It describes the interaction of quarks by means of gluons, which are responsible in the model for the binding of quarks and thus build up hadrons, such as protons and neutrons. Again, symmetry considerations play an important role here. Take the neutron in the nucleus of an atom, which consists of two down quarks and one up quark. If we add up the charges of these quarks $\left(-\frac{1}{3}, \frac{1}{3}, +\frac{2}{3}\right)$, we obtain an electrically neutral particle like the neutron, see

Fig. 6.7. However, quantum chromodynamics would predict for the neutron an unobserved electric dipole moment - a spatial charge separation - which has never been observed in the experiments; a dilemma that needs to be solved. The theorists Roberto Daniele Peccei and Helen Rhoda Arnold Quinn took an approach based on a new symmetry. Their way out was to introduce a new additional symmetry, the Peccei–Quinn symmetry named after them. Since this symmetry is broken in the strong interaction, a new particle can theoretically be observed, in this case a very light, electrically neutral boson. The model that states in particle physics that particles appear when symmetry is broken is the Goldstone Theorem. In the case of the breaking of Peccei–Quinn symmetry in quantum chromodynamics, Frank Anthony Wilczek - an American physicist and Nobel Prize winner - named the new particle after a detergent, Axion, because it would be able to "wash clean" the theoretical description problems of the strong interaction in particle physics. This particle would be very weakly interacting, but would have mass and be stable. If it were to be found - and you will learn in the next chapter that this is what we are looking for - two problems could be solved at once: that of dark matter and that of the theoretical description of *strong interaction*. Does that not sound promising?

The Sterile Neutrino

Besides the WIMPs and the axion, the sterile neutrino represents another hypothetical candidate for *dark matter*. You have already read about neutrinos as possible candidates in this chapter. There are three types of neutrinos in the Standard Model of particle physics: the electron, muon and tau neutrinos, all of which can interact via the weak interaction. Sterile neutrinos represent a hypothetical class of particles that interact gravitationally but otherwise have little interaction with known matter. Again derived from symmetry considerations, their existence would be motivated purely theoretically. Their detection would only be indirectly proven by the statistical frequencies of the different neutrino species. Muon neutrinos could be transformed into sterile neutrinos via the neutrino oscillations, and thus fewer muon neutrinos would be observed compared to the other two types. In 2011, measurements of neutrinos in nuclear reactors showed slight anomalies in the distribution of neutrinos, which may indicate a sterile neutrinos (Mention et al. 2011). Compared to the known neutrinos, the sterile neutrino would be a very massive particle with relatively low velocities, another ideal candidate for our search.

Summary

So we look back on a long era in which dark matter was postulated to maintain the model of gravity on large-scale structures. There are many astrophysical observations and as many interpretations. The time has come to search even more specifically for the nature of dark matter.

After the search for MACHOs in the halo of our galaxy was unsuccessful and researchers came to the conclusion that macroscopic astrophysical objects such as brown dwarfs, stellar black holes or white dwarfs cannot describe the nature of the dark matter phenomenon from a current point of view, physicists are now devoting more attention to the search in the microcosm.

Encouraged by the success of particle physics in the exact description of matter familiar to us and the successful discovery of theoretically postulated particles such as neutrinos, it is particle physics that is currently shaping the search for the nature of dark matter. With numerous experiments looking for candidates from the theoretical extensions of the Standard Model of particle physics, it is hoped to strip dark matter of its Hades cap. How this highly topical research is designing its experiments, what its results so far are and what it is capable of demonstrating so far will be the subject of the next chapter.

References

Mention G, Fechner M, Lasserre Th, Mueller ThA, Lhuillier D, Cribier M, Letourneau A (2011) Reactor Antineutrino Anomaly Phys Rev D 83:073006

Paczyński B (1986) Gravitational Microlensing by the Galactic Halo. ApJ Part 1 304. ISSN 0004–637X

Robert J, et al (2016) A Brown Dwarf Census from the SIMP Survey Accepted for publication in ApJ. https://arxiv.org/abs/1607.06117

7

What is hidden under the Cap of Hades
Current Experiments in the Search for Dark Matter

After the first attempts to reveal the nature of dark matter did not bring the hoped-for success, current experimental physics is pursuing a new approach. A multitude of large and small experiments, all of which have one common denominator: the search outside the familiar canon of the Standard Model of particle physics. Hidden within it is the hope of opening a window to a new era of physics that can reveal the nature of dark matter and, as it were, strip it of its Hades cap - the cap that gave its wearer invisibility in Greek mythology. In order to be able to realise, just like Faust, what holds the world together at its core.

There are two main directions of approaches: the indirect and the direct search for the nature of dark matter. This chapter is devoted to the current state of the art of the methods applied and summarizes the most important research results. You will learn how supposedly assured results can be reassessed in the light of new research results and what we can say about the nature of the "dark" component of the universe from today's perspective. Often the hope for some kind of interaction of dark matter with the forms of matter familiar to us plays a central role, apart from the purely gravitational. If there were such an interaction, we would have a realistic chance of directly measuring dark matter signals in the laboratories of experimental physicists and thus take an important step towards understanding the nature of dark matter. However, we want to focus ourselves first to the indirect search for dark matter.

W. Kapferer, *The Mystery of Dark Matter*, Astronomers' Universe,
https://doi.org/10.1007/978-3-662-62202-5_7

7.1 The Indirect Search for Dark Matter

Currently, two paths are being followed in the indirect search for dark matter. One is the search for signals of a possible interaction of dark matter with itself. The second is the detection of known elementary particles of the Standard Model, which could theoretically occur through dark matter decay processes. There are no narrow limits to the particles and signals that can be expected; rather, it is a broad-range of search for the theoretically conceivable. For example, one looks for charged particles in cosmic radiation, particles like electrons, positrons (the antiparticle of the electron), protons and antiprotons. We are also looking for high-energy gamma rays or neutrinos, which have a different energy distribution than the signals from known astrophysical sources. Figure 7.1 shows schematically what we are trying to observe indirectly: pair annihilations of two weakly interacting particles , the so-called WIMPs. Such a process would produce known particles, which could then be measured.

The problem is that this radiation can also come from known astrophysical sources. It is necessary to be able to distinguish between the sources. This can be done on the one hand via the energy distribution of the radiation, and on the other hand, via the ratios of the different radiation components involved to each other. The hope is to discover a potentially new signal that differs from the known ones and indirectly indicates dark matter.

There are theoretical calculations for indirectly observable signals of WIMP pair annihilation or decay processes and these calculations limit the

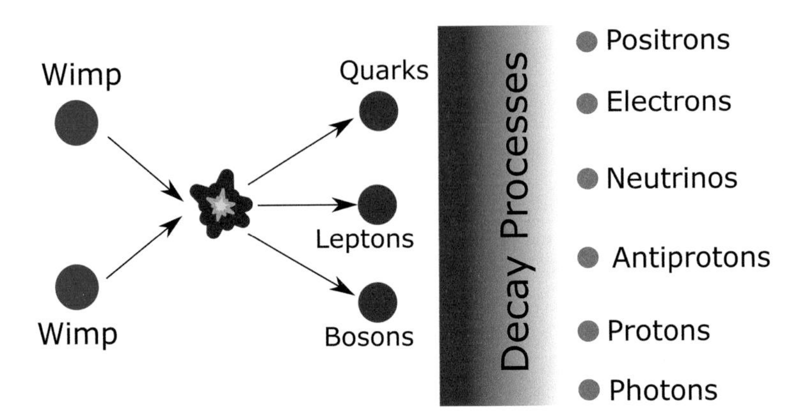

Fig. 7.1 Schematic list of possible particles that can occur in a hypothetical WIMP pair annihilation. Mainly quarks, leptons and bosons would occur. These would then decay into electrons, positrons, protons, antiprotons, neutrinos and radiation, which in turn could be observed

range of signatures to be searched for. However, it must be pointed out that such research results must be interpreted very carefully. Often, as the knowledge about known astrophysical objects progresses, supposedly novel radiation signatures can be explained without the concept of *dark matter*. Let us recall Sir Arthur Eddington who so aptly noted in his 1934 quote that model and experiment should always go hand in hand to provide a convincing model of the world.

The Positron Excess—A Signal from Dark Matter?

We will first look at such a signal from space, which is considered a promising candidate for the indirect detection of dark matter. It is an unexpected excess of a special kind of particles in cosmic radiation.

In the mid-1990s, special detectors on stratospheric balloons were used to study the energies of electrons and their antiparticles, the positrons in cosmic radiation, in greater detail. This radiation of matter and antimatter is caused by high-energy processes in space, such as supernova explosions, black holes at the center of galaxies that attract gas and matter in their vicinity, or neutron stars. In all these cosmic objects, particles are strongly accelerated and dispersed into the depths of space. These initial particles form the so-called "primary cosmic radiation." When these particles hit interstellar matter or the earthly atmosphere, they interact there and produce, for example, neutrons, protons, muons, neutrinos or high-energy gamma radiation. This type of cosmic radiation is called "secondary cosmic radiation."

It was precisely these processes that an American experiment called *HEAT* (High-Energy Antimatter Telescope) investigated in the mid-1990s. It discovered an excess of high-energy positrons in relation to the electrons in the primary cosmic radiation. This was in contradiction to the generally accepted model calculations for cosmic radiation and was a surprise in that it was actually assumed that at higher energies, the number of positrons measured should decrease in relation to the electrons - at least when one starts from typical sources such as supernova explosions or pulsars.[1] Furthermore, a different energy distribution was also found in the spectrum of electrons and positrons, which is interesting insofar as electrons and

[1]Pulsars are a class of neutron stars that rotate quickly and have a magnetic field that deviates from the symmetry of their axis of rotation. As a result, charged particles in the direct vicinity of the pulsar are strongly accelerated.

positrons have the same mass and should be equally influenced by the interstellar magnetic field.

About 10 years later, this behavior of positrons in cosmic radiation was confirmed by the international experiment *PAMELA* (Payload for Antimatter-Matter Exploration and Light-nuclei Astrophysics) on board a Russian satellite. In a period of 2 years (2006–2008), this excess of positrons was detected in the energy range from 1.5 to 100 GeV. The anomaly was then confirmed by an experiment on the International Space Station (ISS), the Alpha Magnetic Spectrometer (short *AMS*) and the *FERMI* Gamma-ray Space Telescope.

So there is an unknown signature in the cosmic radiation that indicates a source that is not yet known. The question was: Could a hypothetical interaction of dark matter with itself be the possible source of the positron excess? WIMPs would be ideal candidates for this as they could hypothetically release matter, antimatter and gamma radiation during a WIMP–WIMP interaction. There are theoretical model calculations for this that predict the expected energy distribution of the observable particles and gamma radiation. Whether this unexpected behavior of positrons in cosmic radiation has its origin in dark matter or in known astrophysical objects is currently the subject of research.

Only recently, Boudaud Mathieu and co-authors (Boudaud et al. 2015) reported that a single fast-rotating neutron star, a pulsar, at a maximum distance of 1kpc and an age of less than 1 million years is sufficient to explain the observed excess of positrons in cosmic radiation. The authors also provide possible candidates, such as the pulsar with the somewhat dry designation J1745-3040. In their calculations, this pulsar would also explain the observations. It should be noted, however, that charged particles are deflected by magnetic fields during their journey to us, making it difficult to clearly assign the particles to an observable source in the sky.

So it remains to be seen whether the physicists have already detected a dark matter signal here or not, since classical objects of astrophysics, such as pulsars, can also be possible causes of the radiation.

Gamma Radiation from the Center of our Galaxy Through WIMPs?

In the model calculations of particle physicists, it is possible that WIMP pair annihilations indirectly release very high-energy electromagnetic radiation - much more energetic than X-rays and thus located in the gamma radiation range. This type of radiation occurs when the strong or weak interaction known to us is involved in conversion processes. In contrast to the charged particles of cosmic radiation, such as electrons or positrons, photons do not feel any deflection by magnetic fields during their journey through the depths of space. This enables us to identify the sources of gamma radiation in the firmament with the help of gamma-ray observatories. Currently, two successful experiments for the detection of gamma radiation from the depths of space are active. One is the *FERMI* Gamma-ray Space Telescope (in orbit since 2008) and the other is the *High Energy Stereoscopic System*, or short H.E.S.S ., a Cherenkov telescope array in the steppes of Namibia, which has been in operation since 2004. The aim of these two experiments is to study the gamma radiation of astrophysical objects. Since such high-energy radiation cannot simply be deflected and collected by means of a mirror, other methods must be used to study it. In general, one makes use of events in detectors or our terrestrial atmosphere that are indirectly caused by gamma radiation.

The FERMI Gamma-ray Space Observatory has been equipped with a special detector that makes such events evaluable. The detector essentially consists of a cuboid of tungsten foils and semiconductor particle detectors. The incoming gamma radiation interacts with the tungsten in such a way that electrons and positrons are released, which are then registered by the particle detectors. The direction of incidence of the gamma rays can be deduced from this. On the other hand, the energy of the electrons and positrons is measured, which indirectly indicates the energy of the causative gamma radiation. The instrument is designed to observe gamma radiation in the energy range from 20 meV to 300 GeV, where M stands for Mega (millions) and G for Giga (billions). The field of view of FERMI is very large. It covers about 20% of the sky at once. The satellite is configured to map the entire sky in about 3 h.

The H.E.S.S . telescope array in Namibia (see Fig. 7.2) pursues a different approach to the measurement of high-energy gamma radiation from space. It uses classical astronomical reflecting telescopes to indirectly detect gamma radiation. It exploits an effect that occurs when particles propagate at speeds faster than the speed of light in a medium. However, since this is lower than the speed of light in a vacuum, no fundamental law of physics is "violated." Gamma radiation from the depths of space causes secondary radiation when entering the earthly atmosphere. Strictly speaking, a whole shower of secondary particles (for example, electrons) accelerated to energies greater than the speed of light in the atmosphere. This is the so-called Cherenkov radiation, named after the Russian physicist Pavel Alexeyevich Cherenkov

Fig. 7.2 The H.E.S.S. array with its four 12-m Cherenkov telescopes and the 28-m large H.E.S.S. II Cherenkov telescope near Table Mountain Gamsberg, Namibia. (Source: H.E.S.S. Collaboration, Clementina Medina)

(1904–1990). This radiation has its maximum in the blue range of visible light and can also be observed as a blue glow in decay pools of nuclear power plants. Astronauts report that from time to time they perceive a blue glow when their eyes are closed. This is because gamma radiation interacts in the liquid of the eyeball, releasing the said Cherenkov radiation inside the eye. What is typically avoided in observational astronomy, namely the influences of the Earth's atmosphere, is here the key to draw conclusions about the energy and direction of incidence of cosmic gamma radiation.

As elegant as this type of detection is, a technical challenge lies in the duration of the events. These are literally flashes of light. The cameras of the H.E.S.S . system must be able to record extremely short events and additionally observe them from different angles. Only in this way is it possible to use the telescopes to trace the spatial course of a flash of light in the sky and indirectly deduce the direction of incidence of gamma radiation from space. The four smaller and the central large reflecting telescope of the facility in Namibia can be seen in Fig. 7.2. In contrast to the classical optical telescopes, the high imaging quality is not of central importance for the Cherenkov telescopes. Rather, the large light-gathering power and the high-speed camera are essential to capture the short flashes of light in the high atmosphere.

After this short excursion into current observational technology, we now want to look at a special gamma-ray signature from the center of our galaxy. Since the first gamma-ray observations, it was clear that many astrophysical objects are sources of this high-energy radiation. These include pulsars, supernovae and the vicinity of black holes. When, after 2 years of FERMI observations, a map of diffuse gamma radiation in the energy range from 80 to 200 GeV was evaluated, some researchers were able to find evidence of gamma rays from the central region of our galaxy (Hooper and Goodenough 2011) that could not be explained by the known sources of this radiation. The spectrum of this radiation in the mass range from 7 to 10 GeV would rather correspond to the hypothetical gamma-ray signature of a WIMP particle pair annihilation. The excitement was great, perhaps dark matter had finally been discovered. Figure 7.3 shows a representation of this gamma-ray signature. You can see the entire sky through the "eyes" of the space telescope and zoom in on the central region of our galaxy. All known gamma-ray sources, such as supernova remnants or pulsars, have been subtracted from this zoom, leaving only the radiation from unknown sources. But this subtraction is not a simple deduction from known values. It requires many elaborate processing steps in data reduction and an accurate model of the detector and its sources of error.

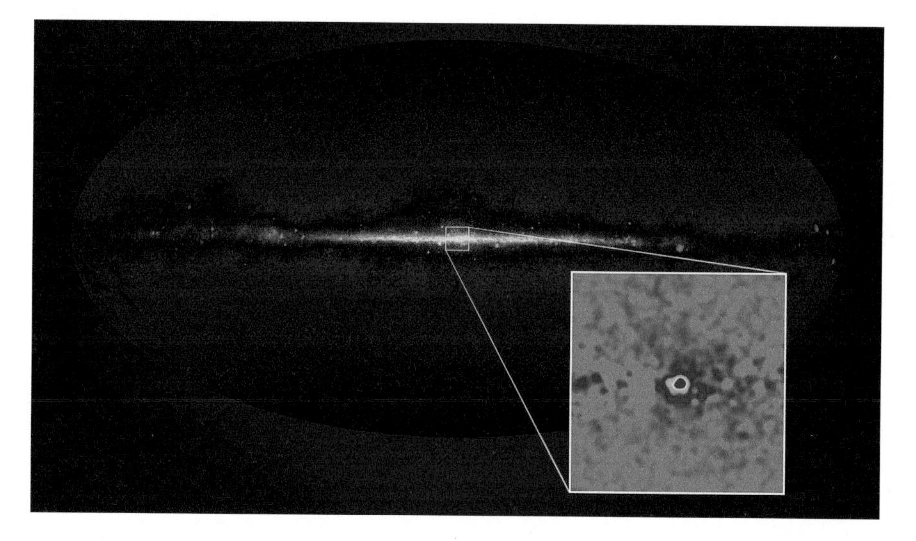

Fig. 7.3 The entire sky in the gamma-ray range as observed by the FERMI Space Telescope. The area outside our galaxy is darkened to focus on the central region of our galaxy. The box is a zoom to the central region. This area has been cleared of all known gamma-ray sources. Nevertheless, an unexpected gamma-ray excess is shown, which could be consistent with the signals of a hypothetical WIMP pair annihilation. Source: NASA/DOE/Fermi LAT Collaboration and T. Linden (University of Chicago)

A few years later, Alex Geringer-Sameth and co-authors (Geringer-Sameth et al. 2015) were able to discover a similar excess of gamma radiation from the dwarf galaxy Reticulum II in the data of the FERMI Space Telescope. According to the authors, the signal was in the energy range of about 2–10 GeV, comparable to the signal from the center of our galaxy. Again, this signal could be well motivated by the radiation of hypothetical WIMP pair annihilations.

In the same year 2015, these discoveries were promptly followed by an analysis of the data by the FERMI and Dark Energy Survey (DES) research consortium (Drlica-Wagner et al. 2015) itself. The research consortium could not confirm the excess of gamma radiation from the central region of these galaxies. The "disappearance" of the gamma-ray excess was due to an improved procedure for data reduction by the consortium. As a result, it was possible to improve the resolution of known astrophysical gamma-ray sources by about 50%. This increase in the detection accuracy of point sources simply made the gamma-ray excess disappear from the data. The supposed WIMP pair annihilation signal actually consisted of unresolved point sources such as pulsars.

Another study by the H.E.S.S . team (Abdalla et al. 2016) could not detect this excess of gamma rays from the center of our Milky Way either. At present, therefore, there seems to be no evidence of dark matter pair annihilations in the center of our galaxy or neighboring dwarf galaxies. So far, all supposedly observed "gamma-ray excesses" have remained unconfirmed due to better data reduction methods and complementary measurement techniques, such as the H.E.S.S . II telescopes. From today's perspective, gamma-ray astronomy does not provide us with a clear picture of the nature of dark matter.

Do Neutrino Observatories See Dark Matter?

In contrast to gamma-ray astronomy, with its search for the signatures of hypothetical WIMP pair annihilations, neutrino observatories follow a different approach for the detection of dark matter. It is the observation of locations where dark matter should be most densely concentrated in our solar environment: in the center of the Earth or the Sun.

The basic idea of this hypothesis is that during the journey of our solar system through the disk of our galaxy dark matter particles would be captured by the planets and the Sun. Just as if our Earth or the Sun, or any other object in our solar system, were a "vacuum cleaner" for dark matter.

Why vacuum cleaners? A few of the hypothetical WIMP particles could collide with the atomic nuclei of the Earth or Sun and release some of their energy, provided there is an interaction between dark matter and known matter. Thus, WIMPs would be slowed down and would not be able to escape from the gravitational field of the Sun or the planets. Statistically, this happens most often where there is a high density of matter, i.e. in the interior of the Earth or the Sun. Over billions of years, the concentration of dark matter there would continuously increase until there is enough of it to cause enough WIMP pair annihilations to produce a detectable signal.

And this is exactly where the neutrino plays an important role as a possible decay channel of WIMP pair annihilation. Theoretical calculations predict a neutrino signal depending on the energy and mass of the WIMP particles and the structure of matter in the Earth's or Sun's interior, which could be observed with the detectors of IceCube (see Fig. 6.11). The IceCube Neutrino Observatory is located near the South Pole and collects data under the ice at depths ranging from 1,450 to 2,450 m. The facility covers an ice volume of 1 km^3.

However, the current results for such a dark matter signal from the interior of our Earth are sobering. Over a total observation period of 327 days in the years 2011/2012, no signal could be found that emerged from the known neutrino background (Aartsen et al. 2017a). The search for signals from the interior of the Sun also remained unsuccessful until the end of 2016 (Aartsen et al. 2017b). Here "unsuccessful" should not mean "in vain" because these measurements made it possible to exclude many model calculations on hypothetical dark matter particles. Possible supersymmetric models could also be rejected due to missing measurement events. It should not be forgotten that the experiment has been very successful in the study of neutrino oscillation. All in all, this research places valuable limits on the theorists' models.

7.2 The Direct Search for Dark Matter

In contrast to the indirect search, the direct search is dedicated to the detection of an interaction of dark matter with ordinary matter directly in the physicists' laboratories. In order to ensure the best possible shielding of interfering signals, they are usually located deep underground, for example, in disused mines. Some of these experiments were indeed able to record a number of unusual signals. The next section will summarize the methods used and the results obtained so far. Starting from smaller feasibility studies, the trend - as in many areas of current cutting-edge research - is towards ever larger facilities, operated by large international collaborations.

Many Experiments—One Goal

A complete overview of all experiments for direct detection of dark matter would go beyond the scope of this book. Therefore, a selection of the "most successful" experiments was made, although such an evaluation can certainly not be objective. A few important representatives from the different detector classes are presented. A list of experiments and links to their websites can be found at https://lpsc.in2p3.fr/mayet/dm.php. There are currently 52 experiments listed, which shows that the direct search for dark matter is very active.

All these experiments have to cope with a common problem: the differentiation of the signals of dark matter from those caused by other, conventional interactions. This requires a very good knowledge of the complex detectors in order to avoid misinterpretations of initially "inexplicable" signals, as we have seen with the FERMI Space Telescope.

The original ideas of many experiments in this field of research date back to the mid-1980s and were first published by physicists such as Andrzej Drukier, Leo Stodolsky, Piere Sikivie, Mark Goodman and Edward Witten. These works describe experiments that could essentially prove possible interactions of dark matter candidates with matter familiar to us.

Let us take, for example, the idea of Piere Sikivie to prove the hypothetical axions of 1983 (Sikivie 1983). He showed that one could stimulate the decay of axions by an electromagnetic field configured in a certain way. In this process, axions would then be transformed into measurable microwave radiation. However, the resulting signal would be extremely weak, in the order of 10^{-22} watts and would have to be amplified accordingly for a measurement - a nontrivial challenge for technology.

For the WIMPs, the early experimental approaches were developed by Mark Goodman and Edward Witten in their 1985 publication (Goodman and Witten 1985). Their approach is that dark matter particles could have a small but measurable interaction with ordinary matter. The WIMP would with a certain probability collide with a core of the detector element and thus release energy to the detector. The detector material would consist of superconducting materials measuring a few thousandths of a millimeter, which are encased in a nonsuperconducting shell. The entire experimental setup is exposed to a strong magnetic field.

Superconducting materials are characterized by the fact that they can conduct electrical currents without resistance. This property generally occurs at very low temperatures. Interestingly, the superconducting property changes abruptly with increasing temperature. Like a switch, the effect of superconductivity can be switched on or off by exceeding or falling below a certain temperature. And this is exactly what is used for the detection of WIMPs.

When WIMPs encounter superconducting particles, they would be heated. Such a superconducting measuring particle could suddenly lose its superconducting ability and become electrically resistant. The magnetic field applied to the experiment would be suddenly changed and one would be able to measure a signal of the interaction. According to Goodman and Witten, a detector of 1 kg detector mass would measure more than one event per day according to many of today's model calculations for WIMP interactions. Many hypothetical WIMP candidates could thus be excluded in the absence of detection. The main problem with this class of experiments are background signals caused by natural radioactivity or cosmic radiation. A sophisticated system is needed to detect these signals and distinguish them from possible WIMP interactions.

We will now take a closer look at some direct search experiments and their results so far.

Darkness that Warms

The EDELWEISS Experiment

One representative of the experiment proposed by Goodman and Witten for the detection of WIMPs is the EDELWEISS experiment, which is operated by an international collaboration of French, German, Russian and British researchers. To realize such an experiment, the choice of the location is of central importance. One needs a location that minimizes all external influences. No high natural radioactivity, no large temperature fluctuations, no electromagnetic fields in the environment, well shielded against cosmic radiation, are just some of the points to be considered. For the EDELWEISS experiment, it was therefore decided to use an underground laboratory. It was installed in the laboratories in the *Souterrain de Modane*, between Modane and Bardonecchia in the French-Italian border area. The site is located at a depth of around 1,700 m under Mount Frejus in a branch of the Mont Cenis Tunnel and is ideally shielded against all kinds of external influences.

The detector of the experiment consists mainly of cells of germanium crystals. These are cooled down to about 18 thousandths of a Kelvin and constantly maintained at this temperature. The experiment thus takes place very close to absolute zero. Thermal energies and electron flows are measured in the detector, which is well shielded. Figure 7.4 shows the detector still

Fig. 7.4 The experimental setup of the EDELWEISS detector. The picture was taken during the setup in February 2014. On the right side, you can see a copper cylinder being put over the apparatus. The whole experiment is then packed into a 40-tonne lead jacket (can be seen in gray on the left and right edges). The entire detector is cooled down to 18 thousandths of a Kelvin above absolute zero. Courtesy of the EDELWEISS-Collaboration

open. You can see hockey puck-sized individual detector cylinders, each of which weighs about 800 g. The entire apparatus is shielded by a 20-cm thick lead jacket during the measurements.

On the one hand, temperature differences down to a few millionths of a degree are measured, and on the other hand, electrons released by impacts of particles on the atoms within the crystal are measured. This approach has now been pursued for around 20 years with a wide variety of detectors in these underground laboratories. One tries to prove the WIMPs with ever higher measurement accuracies, for example, with more mass in the detector crystals to increase the possible interaction rates, and an ever better understanding of the background signals.

This increased the accuracy of the measurements by a factor of 40 in the last series of experiments between July 2014 and April 2015.[2]

[2]Silvia Scorza for the EDELWEISS Collaboration; Scorza (2016).

Has the WIMP been proven to be existent in the meantime? No. Unfortunately, no unambiguous dark matter signals could be found. All signals could be explained by other influences such as natural radioactivity and processes within the measuring apparatus. But the absence of hoped-for WIMP signals should also be considered a success as it helps theorists with their model calculations and falsifies many WIMP hypotheses. It remains to be seen whether, with more detectors, longer test series and a growing understanding of other background signals, a clear signal from the WIMPs can one day be detected.

The CRESST Experiment

A similar concept is followed by the CRESST experiment (Cryogenic Rare Event Search with Superconducting Thermometers), an international collaboration of the German Max Planck Institute for Physics (Munich), the Universities of Munich and Tübingen, the University of Oxford, the Austrian Institute for High Energy Physics and the Italian INFN (Istituto Nazionale di Fisica Nucleare; National Institute of Nuclear Physics). Again, a possible interaction of a WIMP with the atomic nuclei of a crystal is assumed as working hypothesis. The energy transferred to the system is measured by very sensitive temperature sensors. Similar to the EDELWEISS experiment, CRESST is shielded from external influences in a laboratory of the INFN in a research hall in the Gran Sasso tunnel system in Italy. The detector in its current design consists of calcium tungstate crystals ($CaWO_4$), which are produced in high purity at the Technical University of Munich. As suggested in the 1985 work by Goodman and Witten, superconducting materials that change their electrical resistance significantly with the slightest temperature changes act as temperature gauges. In addition, a light detector in this experiment measures possible light flashes that can occur with typical background signals such as natural radioactive radiation or electrons. This allows to distinguish possible WIMP signals better from ordinary background signals. Figure 7.5 shows the installation of the detector. The detector is suspended from a thermally highly stable heat bath, which is located above the apparatus and keeps the measuring apparatus at a constantly low temperature.

After measurements between June 2009 and April 2011, the consortium researchers reported unexplained signals in the detector that could not be explained by known background events such as natural radioactivity. As a possible cause for these unexpected signals, the authors published impacts

Fig. 7.5 The optimized CRESST detector during installation for the current measurement series, which started in summer 2016. Courtesy of A. Eckert/MPP

of WIMPs on the atomic nuclei (Angloher et al. 2012) of the detector. The possible mass range of the WIMPs would be in the range of 25.3 and 22.6 GeV according to the measurement data. They would therefore be lighter representatives of their class. In total, about 50 such events could be recorded, which indicate a WIMP interaction in the detector.

Was this finally the first human-measured interaction of the dark component of the cosmos with the world familiar to us?

Two years later, the research consortium published new results of their research. This time, the data was derived from an improved experimental setup. The authors wrote in the original paper: "A possible excess over background discussed for the previous CRESST-II phase 1 (from 2009 to 2011) is not confirmed" (Angloher et al. 2014). Thus, the signals of the first measurement series could no longer be reproduced. Just as in the EDELWEISS experiment, the CRESST experiment did not deliver any dark matter signals to date. However, it is possible that in the future improved detectors and higher detector masses combined with higher interaction probabilities will clearly prove WIMP interactions.

The Four Seasons of WIMP Detection

The DAMA/LIBRA Experiment

A further experiment in the field of the direct search for dark matter is the DAMA/LIBRA experiment, whereby DAMA (DArk MAtter) as an umbrella project combines several experiments for the direct search for dark matter. Just like CRESST, these experiments are located in the tunnels of Gran Sasso in Italy and, according to the official website, form an observatory for the discovery of rare processes. Similar to the CRESST experiment, WIMP impacts on crystal atomic nuclei release heat and electrons, which can be measured subsequently.

The LIBRA experiment, abbreviated for *Large sodium Iodide Bulk for RAre processes*, has been in operation for more than 10 years and continues the predecessor experiment from the 1990s, which simply bore the name of the detector material: NaI (Sodium Iodide doped with Thallium). It is remarkable that DAMA/NaI was the first such experiment to publish a clear WIMP signal as early as 1998 (Bernabei et al. 1998).

We want to take a closer look at these WIMP signals. First of all, as always, there is the question of how other signals in the detector, such as those caused by natural radioactivity, can be distinguished from WIMP signatures. What characteristics did the WIMP signal have that it was interpreted as such? The answer is provided by a theoretical work from the mid-1980s by Andrzej Drukier, Katherine Freese and David Spergel (Drukier et al. 1986). This publication describes a detector - similar to the detectors presented so far - and then goes into the expected background signals in detail. The authors also describe a very special signal that only WIMPs would produce, a signal that should change in the course of a year. In Fig. 7.6, the motivation for such an annual fluctuation of the WIMP interactions in the detectors is outlined.

While the Sun orbits around the galactic center at around a few hundred km/s, the Earth moves around the Sun on its orbit. However, the solar

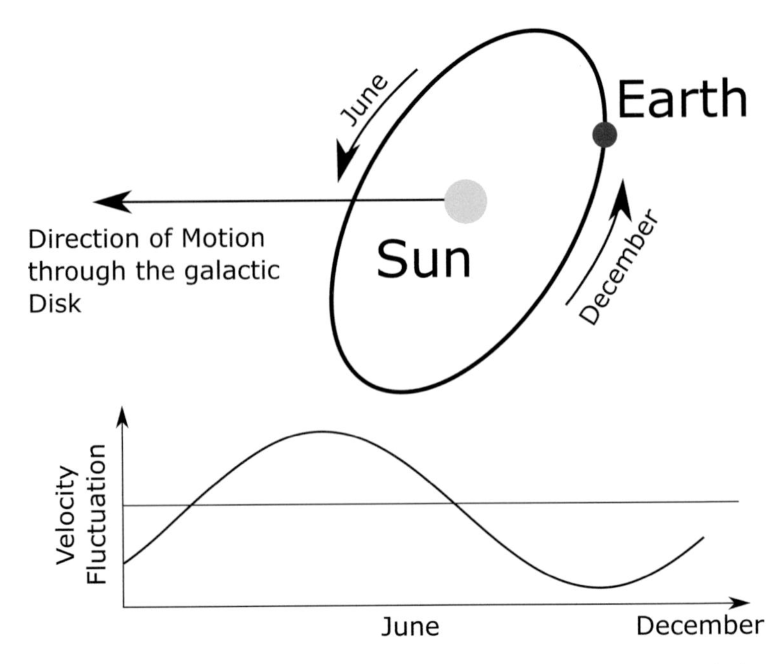

Fig. 7.6 The annual fluctuation of the Earth's orbital velocity around the galactic center. The Sun moves with constant velocity around the center of the galaxy. Because of the orbit of the Earth around the Sun, which is inclined at about 60° relative to the galactic plane, the galactic orbital velocity of the Earth varies. Thus, it is somewhat faster in summer than in winter - depending on whether the Earth moves along its orbit around the Sun in the direction of rotation around the galactic center or in the opposite direction. This fluctuation of orbital velocity also means that within 1 year, the Earth moves sometimes faster and sometimes slower in relation to the particles of dark matter

system is inclined by around 60° to the galactic disk, which means that the Earth orbits the galactic center a little faster than the Sun in summer and slower in winter - depending on whether the Earth moves along its orbit around the Sun in the direction of rotation around the galactic center or in the opposite direction. In the model calculations of dark matter, it rests on average relative to the stars, the planets or the interstellar gas within a galaxy. If this corresponds to reality, this would lead to higher impact energies in the detectors of the direct search for dark matter, which basically measure the impact energy of the WIMP particles with the detector crystals, in summer than in winter. Put simply, the Earth collides with the hypothetical, omnipresent WIMP particles somewhat faster in summer than in winter. If one were to measure such a fluctuating signal, this would be a strong indication

Fig. 7.7 The exposed copper jacket of the DAMA/LIBRA experiment during maintenance work. You can also see the lead-colored, outer jacket of the detector for shielding from external influences. Courtesy of the DAMA Collaboration

of WIMPs as the causal partner of the interaction. Natural radioactivity or cosmic radiation would not exhibit such periodicity.

And it is precisely this annual variation in the energy of events that both DAMA/NaI and DAMA/LIBRA were able to discover. Figure 7.7 shows the LIBRA detector packed in copper during maintenance work. Like the EDELWEISS detector, the experiment is cooled and embedded in a copper jacket. The whole experiment is then again packed in a lead sheath deep inside the mountain for the best possible shielding.

So let us note: In 1998, the DAMA research team published measurement data that cyclically changed according to the Earth's orbit around the Sun, exactly as described by Andrzej Drukier, Katherine Freese and David Spergel in their work from the 1980s. Figure 7.8 shows the measurement curve mentioned above. It covers the period from 1995 to 2009 and shows the much smaller measurement errors in the blue area of the measurement data, which are due to the new DAMA/LIBRA detector. The plotted continuous curve, which represents an annual fluctuation of the data, approximates the temporal fluctuation of the data as closely as possible. It has a period of 364.6 ± 1 day, almost exactly 1 year, as theoretically predicted; a strong indication of WIMP-particle interaction.

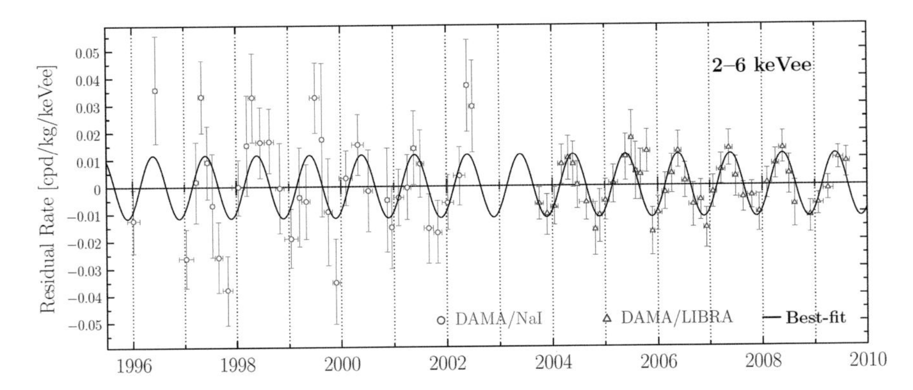

Fig. 7.8 The measurement of the cyclic change in the mean energy of the detector events of DAMA/NaI (1995–2002) and DAMA/LIBRA (2003–2009). For this periodic approximation curve, a period duration of 0.999 ± 0.002 years was found. Source: Freese et al. (2013)

But could other experiments also measure this concise periodic change in the measurement data?

The CoGeNT Experiment

In addition to DAMA/NaI and DAMA/LIBRA, there is another experiment that was able to detect this annual fluctuation in its data. It is CoGeNT, a germanium detector in the Soudan Underground Laboratory of the University of Minnesota (Universität Minnesota), USA. This experiment is also located in a disused iron mine at a depth of around 713 m to seal off external influences. In Fig. 7.9, you can see the open experiment. Once again, a lead sheath is used to isolate the detectors from the environment. The whole mode of operation of the experimental setup is very similar to the EDELWEISS experiment. Between 2009 and 2011, the detector, which weighs around 440 g, was able to measure a number of events. The CoGeNT collaboration published its measurement results in 2011 (Aalseth et al. 2011). Their measurements also showed an annually fluctuating behavior, similar to the DAMA/LIBRA experiment. The periodic time found with CoGeNT was 347 ± 29 days, again approximately the period duration of 1 year.

Is the WIMP and thus Dark Matter Now Discovered?

Currently data from two experiments are available that could be well explained with hypothetical WIMP particles in our Milky Way. But unfortunately,

Fig. 7.9 The CoGeNT detector during installation. You can see the lead sheath that protects the cooled detector from external influences in a 713-m deep iron mine. Courtesy of Juan Collar; https://cogent.pnnl.gov/gallery.stm

neither EDELWEISS nor CRESST observed such a signal. Also other similar experiments to search for WIMPs, like the CDMS (Cryogenic Dark Matter Search) in the Soudan Underground Laboratory, the Super Kamiokande in Japan, the Russian Baksan experiment or the experiments XENON 10 and XENON 100 in Gran Sasso could not confirm the existence of WIMP interactions until the end of 2016. Currently, the SABRE experiment (Sodium iodide with Active Background REjection) is being set up in an Australian gold mine, which is technically similar to the DAMA/NaI experiment. On the one hand, the DAMA measurement series are to be confirmed, and on the other hand, due to its location in the southern hemisphere, seasonal effects of the northern hemisphere, such as temperature fluctuations, are to be excluded. One tries to discover the same measurement curve as in DAMA/NaI.

All in all the search for the WIMPs remains open and exciting. Better detectors will hopefully be able to clarify in the coming decades the question whether WIMPs exist or not. From today's point of view, WIMP particles are unfortunately still of a hypothetical nature and cannot be used to explain the nature of dark matter.

Dark matter in the Particle Accelerator

The ATLAS and CMS Detectors at CERN

WIMPs are also actively searched in two of the huge detectors of the particle accelerator of the Conseil Européen pour la Recherche Nucléaire (CERN, European Council for Nuclear Research), the ATLAS (A Toroidal LHC

ApparatuS) and the CMS (Compact Muon Solenoid). In the elementary particle accelerators at CERN, protons are accelerated to almost the speed of light and brought to collision. The collisions, which occur inside huge particle detectors, are the starting point for the actual experiment. One such detector is shown in Fig. 7.10. ATLAS is a hall-filling machine, consisting mainly of magnets and various semiconductor detectors for measuring the particles created by the collision. The detectors are able to measure the charged particles and photons of the Standard Model that are produced as a result of the collisions.

Again a conservation law plays the central role here, this time the conservation of momentum, that is, the conservation of that physical quantity that can be assigned to a body by its mass and velocity. When two protons collide, new particles are created, which are measured in the detector. These particles fly with different directions and speeds, i.e. momentums, through the detector. Before the collision, there was no momentum transverse to the proton beam because the momentum before the collision of the particles is aligned exactly along the center of the detector. After the collision, the momentums of all observed particles are measured, which are created by the collision. These momemtums have a directional component that does not only point along the particle beam. Rather, the momentums of the new particles are distributed in all possible directions. If you now add up all the measured momentums, the total momentum must be perpendicular to the particle beam and thus to be zero. If this momentum is not zero, it is assumed that a new type of particle, invisible to the detector, has been created. This particle has probably carried the momentum away from the system and left us with an alleged violation of the law of conservation of momentum.

As of November 2016, no indications - beyond the inaccuracy of the detectors involved - of "violations" of the conservation of momentum

Fig. 7.10 View into the interior of the ATLAS detector at CERN (November 2005). Courtesy of Frank Hommes

principle could be found. However, due to this non-detection, mass areas for WIMPs can be excluded, which is of great importance for further research. By increasing the collision energy of the protons at CERN, it may soon be possible to open a door to physics beyond the Standard Model and perhaps discover one or two candidates for dark matter.

Summary

For about 20 years, there has been an intensive search for dark matter signatures, both directly and indirectly. Many different experiments have been developed and conducted for this purpose. After initial indirect signs of dark matter due to gamma radiation from the center of the Milky Way, better data reduction methods have led to the conclusion that these signatures may also originate from known astrophysical objects, such as pulsars or supernova remnants. For example, the excess in gamma radiation found in the data from the FERMI Space Telescope at the centers of both the Milky Way and nearby dwarf galaxies could not be verified using H.E.S.S . II. In addition, with better data-reduction algorithms, these signatures were no longer reproduced in the data from FERMI itself.

The direct search for a dark matter signal in the excellently shielded laboratories deep underground is an ambiguous situation. Probably direct search experiments such as DAMA/NaI, DAMA/LIBRA and CoGeNT have been able to detect possible signals from WIMPs. However, other similar experiments were not successful. Here one must wait for further data and new experimental instruments. Perhaps an Australian experiment, very similar to the DAMA/LIBRA setup, can provide clarity in the next few years. Even the largest experiment ever conducted by particle physicists to date at CERN has not been able to find any evidence of dark matter.

However, by means of more refined measurement techniques, increased detector masses and therefore higher interaction probabilities, the search for dark matter will be pursued further and perhaps - with a bit of luck - be successful.

References

Aalseth CE, Barbeau PS, Colaresi J, Collar JI, Diaz Leon J, Fast JE, Fields N, Hossbach TW, Keillor ME, Kephart JD, Knecht A, Marino MG, Miley HS, Miller ML, Orrell JL, Radford DC, Wilkerson JF, Yocum KM (2011) Search for an annual modulation in a p-type point contact germanium dark matter detector. Phys Rev Lett 107:141301

Aartsen MG et al (2017a) First search for dark matter annihilations in the Earth with the IceCube detector. Eur Phys J C 77(2, article id. 82):11

Aartsen MG et al (2017b) Search for annihilating dark matter in the Sun with 3 years of IceCube data. Eur Phys J C 77(3, article id. 146):12

Abdalla H et al (H.E.S.S. Collaboration) (2016) H.E.S.S. Limits on Linelike Dark Matter Signatures in the 100 GeV to 2 TeV Energy Range Close to the Galactic Center Phys Rev Lett 117:151302

Angloher G, Bauer M, Bavykina I, et al (2012) Results from 730 kg days of the CRESST-II Dark Matter search Eur Phys J C 72:1971

Angloher G, Bento A, Bucci C et al (2014) Results on Low Mass WIMPs Using an Upgraded CRESST-II Detector Eur Phys J C 74:3184

Bernabei R et al (1998) Searching for WIMPs by the annual modulation signature. Phys Lett B 424(1–2):195–201

Boudaud M et al (2015) A new look at the cosmic ray positron fraction. a&A 575:19

Drlica-Wagner A et al (2015) Search for gamma-ray emission from DES dwarf spheroidal galaxy candidates with Fermi-LAT Data. ApJL 809:L4

Drukier AK, Freese K, Spergel DN (1986) Detecting cold dark-matter candidates. Phys Rev D (Particles and Fields) 33(12):3495–3508

Freese K et al (2013) Colloquium: Annual modulation of dark matter. RevPhys 85(4):1561–1581

Geringer-Sameth A et al (2015) Indication of Gamma-Ray Emission from the Newly Discovered Dwarf Galaxy Reticulum II. Phys Rev Lett 115(081101):1–6

Goodman M, Witten E (1985) Detectability of certain dark-matter candidates. Phys Rev D (Particles and Fields) 31(12):3059–3063

Hooper D, Goodenough L (2011) Dark matter annihilation in the Galactic Center as seen by the Fermi Gamma Ray Space Telescope. Phys Lett B 697:412–428

Sikivie P (1983) Experimental tests of the „invisible“ axion. Phys Rev Lett 51(16):1415–1417

Scorza S (2016) EDELWEISS-III experiment: Status and first low WIMP mass results. J Phys Conf Ser 718:1–6

8

Concluding Remarks

For about a hundred years, there has been no satisfying answer to the question about the nature of the matter that dominates the universe. The mystery arose when the masses derived from the dynamics of galaxies were compared with those derived from the luminosity of their objects. In 1922, Sir James Hopwood Jeans published his study of the distribution of stars in the Milky Way, in which he wrote: "…there must be about three dark stars in the universe to every bright star." This statement was essentially similar to Friedrich Bessel's postulate of the dark companion star of the Sirius binary star system from 1844. Some 20 years later, it was technically possible to observe Friedrich Bessel's postulated companion star, a technical masterpiece achieved by Alvan Clark. But it did not stop with the work of Hopwood Jeans. Further signs of this dark component were discovered, for example, by Horace Babcock in 1939, who showed that the dynamics of the Andromeda galaxy cannot be reconciled with its observed mass distribution. Twenty years later, in the early 1960s, Vera Cooper Rubin and her colleagues were able to reproduce exactly this behavior for numerous galaxies. As a result of this work the scientific community began to accept that a significant amount of dark matter in the galaxies is indeed required to reconcile their observed rotation with the laws of gravity.

On even larger scales, the galaxy clusters, Fritz Zwicky was able to demonstrate in 1933 by a dynamic mass estimate of the Coma galaxy cluster that an even greater mass discrepancy exists. It was also he who coined the term dark matter in a work in 1933. In his publication about the present knwoledge on the observation of redshifts in extragalactic nebulae, he

W. Kapferer, *The Mystery of Dark Matter*, Astronomers' Universe,
https://doi.org/10.1007/978-3-662-62202-5_8

remarked on the scattering of velocities in the Coma cluster of galaxies that observed scattering would probably require an average density of matter in this galaxy cluster at least 200 times higher than observed to be consistent with the theory of dynamics.

With the appearance of new observation windows in the electromagnetic spectrum, such as satellite-based X-ray astronomy, about four times more ordinary matter in the form of hot gas was discovered between the galaxies in galaxy clusters. But to explain the physical properties of this so-called intracluster gas, it is again necessary to have an addititional mass, i.e. dark matter.

When, in the following decades, observations of the cosmic microwave background and distant supernovae also called for dark matter in the universe in order to reconcile theoretical models with observations, the time was ripe for a targeted search for the nature of dark matter. No easy task, because what should one be looking for?

One approach was to watch out for massive halo objects in our Milky Way (MACHOs) using general relativity and the gravitational lensing effect described therein. Unfortunately, to this day, it is still not possible to detect a sufficient number of MACHOs around our Milky Way. Therefore, such objects are currently ruled out as candidates for dark matter. Even the search in the microcosm, in the experiments of particle physicists, has so far been unable to clearly reveal the nature of Dark Matter. At present, theoretical extensions of the Standard Model, such as supersymmetry and the hypothetical massive particles it contains, commonly referred to as WIMPs, are the most promising candidates for Dark Matter, along with the axions and sterile neutrinos. However, there is no clear experimental proof of these hypothetical particles. Although two experiments, DAMA/NaI, DAMA/LIBRA and CoGeNT have already reported a positive detection of WIMP-like particles, these observations are contradicted by numerous non-observations of other experiments, experiments such as EDELWEISS, CRESST, CDMS and XENON, to name only the most important ones. The direct detection of a dark matter particle has not yet been successful.

But after all, what speaks for the broad acceptance of dark matter in the research community? Not least the success of the standard theory for structure formation in the cosmos, the Λ-CDM model. Using sophisticated simulations on powerful computer systems, physicists are now able to reconstruct the formation of galaxies, galaxy clusters and the cosmic network of matter in detail. A statistical comparison of the results of cosmological simulations with observations agrees very well over wide ranges. And since dark matter plays a central role in all these simulations, it is difficult to escape this concept.

Nevertheless, attempts have been and are being made to explain the observations without any dark matter whatsoever by modifying the theory of dynamics. Such theories have so far been able to describe rotation curves on galactic scales completely without dark matter. But on large scales, these approaches also require dark matter, albeit a much smaller amount. This approach is probably not experiencing a large impact in the research community because such modifications have no causality in an understood physical process and thus represent a purely mathematical fitting-approach to observed quantities. An approach that requires a great deal of fine-tuning in its parameters and ultimately does not completely succeed without additional, unobservable dark matter on large scales. This is very much in line with Ockham's razor, which states that of several models for an observation, the one with the fewest variables and assumptions should be used.

Perhaps future observations, perhaps the emerging observational gravitational wave physics or new experiments by particle physicists, will reveal to us the nature of dark matter. Chance certainly plays an important role in this. At the end of 2016, Ray P. Norris, an Australian astronomer, published that of the ten greatest discoveries made with the Hubble Space Telescope, only one was planned as a key observation in the designing phase of the telescope (Ray 2016). (These were observations of Cepheid stars, which helped to better determine the Hubble constant.)

Thus, we are now at the beginning of the twenty-first century at a point where it is clear that we have a fundamental problem of understanding the world on galactic and even larger scales. Either there is a dominant mass of dark matter in the vastness of space or our current theory of gravitation fails on large scales. Either way, it will remain exciting in the field of dark matter research. Future experiments or possibly more general models of the dynamics of masses will perhaps one day make the problem of dark matter seem like the ether theory, a theory that has led natural scientists to search in vain for a medium for the propagation of light. In the end, the ether was not discovered, but revolutionary new ideas came up, like the theory of relativity or quantum mechanics. Ultimately, these concepts had such a far-reaching impact that their technical applications are still revolutionizing our everyday life. We can be curious about what lies behind the mystery of dark matter.

References

Ray PN (2016) Discovering the Unexpected in Astronomical Survey Data. Publ Astron Soc Aust 34(id.e007):10

Index

A

B

C

W. Kapferer, *The Mystery of Dark Matter*, Astronomers' Universe,
https://doi.org/10.1007/978-3-662-62202-5

Printed in the United States
by Baker & Taylor Publisher Services